South University Library
Richmond Campus
2151 Old Brick Road
Glen Allen, Va 23060

NOV 3 0 2010

Environmental Communication

Second Edition

Richard R. Jurin • Donny Roush
Jeff Danter

Environmental Communication

Skills and Principles for Natural Resource Managers, Scientists, and Engineers

Second Edition

 Springer

Richard R. Jurin
School of Biological Sciences
University of Northern Colorado
Greeley, CO
USA
Richard.jurin@unco.edu

Donny Roush
Odyssey School
Denver, CO
USA
donnyroush@gmail.com

Jeff Danter
The Nature Conservancy
Florida Chapter
USA
jdanter@tnc.org

ISBN 978-90-481-3986-6   e-ISBN 978-90-481-3987-3
DOI 10.1007/978-90-481-3987-3
Springer Dordrecht Heidelberg London New York

Library of Congress Control Number: 2010924748

© Springer Science+Business Media B.V. 2010
No part of this work may be reproduced, stored in a retrieval system, or transmitted in any form or by any means, electronic, mechanical, photocopying, microfilming, recording or otherwise, without written permission from the Publisher, with the exception of any material supplied specifically for the purpose of being entered and executed on a computer system, for exclusive use by the purchaser of the work.

Cover illustration: 'Rock Wall in Forest' by Ken Shearer

Printed on acid-free paper

Springer is part of Springer Science+Business Media (www.springer.com)

Preface

A rock wall stands on a forest floor in New England. This wall was built when the United States was a young nation, by human hands. Between its origin (circa 1800) and now, there have been shifts in this scene – some marked, some less apparent.

On both sides of the wall, a second-growth forest stands, having replaced virgin woods, removed in all likelihood well before the fence was put in place. On the boulders comprising the fence, lichen colonies have formed, a phenomenon requiring decades without disturbance. This structure likely used to mark the edge of a farm field, from which food and fiber were extracted. The practical purposes of the wall were to mark a boundary and to provide a place for disposal of rocks, to lessen the difficulty of plowing. Such rock fences are the residue of difficult, extensive work.

The wall's communicative purpose was to proclaim dominion. Pioneers who built rock fences meant to let others know that this was their land, that a family made its living here. Theirs was a thoroughly un-modern and close relationship with the land. The wall is a message, encoding a bit of human–land interaction in the United States. It is an American environmental communication.

Photographer Ken Shearer captured this wall one autumn while touring the backwoods of Vermont and Maine. He recalls miles and miles of these rock walls standing along country lanes. Shearer uses large format cameras, to produce photographs of tremendous beauty and power. We liked his photograph very much and hope you do too.

This book is about communication and our environment. We mean for our cover to be indicative of what is inside.

We hope to convey additional messages through the graphic design of the cover and this page. First, the earthy tones in the cover photograph are meant to remind viewers of the natural world. The greens, yellows, browns and grays within the image denote autumn in a temperate, deciduous forest, an ecosystem familiar – at least vaguely – to most. One shade of gray is picked up and repeated as the drop capital for this preface. Here our meaning was subtler. Environmental communication deals in many, many gray areas. Uncertainty is rife in environmental science; rarely do we have clear answers about the environmental effects of post-industrial human society. Most always, clear answers take much longer to uncover than we wish. Message-makers have to consider what is known and present remedies based on the preponderance of the evidence, evidence usually still clouded by at least shadows of reasonable doubts. This task is not easy.

But, it is a necessity if we – as individuals, as groups, and as a society – are to continue to redirect ourselves toward sustainability. We are not alone in this type of thinking about environmental communication. Indeed, voices of leading scientists, communicators, and activists have grown more voluminous and more harmonious, breaking into a distinct chorus in the 1960s. Since the turn of the millennium, there's been a notable amplification as environmental messages have accelerated, become heightened, and have penetrated further into our collective psyche. Behaviors are shifting as a result. Green is in.

To keep society moving toward sustainability, environmental communicators have to work skillfully with more than mere information. 'Conservationists, who after all are inspired to work in conservation not just because of information they have but because of how it makes them *feel*, need to find better ways to fuse scientific information with a wider array of human values,' writes Safina (2008) in *A Passion for This Earth: Writers, Scientists, and Activists Explore Our Relationship with Nature and the Environment*. 'Think of it as a question of *translation*.

'This is not a matter of manipulating people; it's a matter of communicating matters of the utmost importance to the future of life. Science needs a human face. We need to increasingly focus on formulating messages that allow people to recognize themselves.

'Because we are talking about health and the ability of our planet to continue supporting life, formulating such messages should be possible. There has never been a more important time to take a new approach to communicating.' [Emphasis in original.]

We wholeheartedly agree. Indeed, such environmental communications – defensible in their factual claims, voiced clearly by reputable professionals, and prescriptive at their conclusions – are imperative and need to be applied by all professionals involved in the management of the Earth's natural resources. Scientists, engineers, and natural resource managers must repeatedly bring knowledge and solvent ideas to bear on our environmental problems, as part of larger public discourses. Their messages need to be salient and sensible, as the wider populace will not necessarily accept as stone-carved fact, or even respect, the contentions of trained environmental professionals based solely on reputable standing. In our twenty-first century world, mediated messages exist in a cluttered marketplace of ideas; much is plain noise and distracting. Serious dialogue about our environment must cut through this clamor. In this book, we hope to provide pointers to assist in the construction of well-planned, well-targeted, and efficacious environmental communications.

So, this work is meant as a compendium for anyone who communicates about nature, natural resources, ecology, and environmental processes. Whether you work for a non-profit organization, as part of a for-profit business, within a government agency charged with management and policy-making, or as a free-agent activist, we aim to provide a handy reference filled with focused suggestions and know-how. The high stakes of environmental situations demand more from one's communications.

In your use of this book, realize that even as we discuss communications in discrete 'chunks,' we attempt to point out threads of integration among topics. We always seek to retain a holistic perspective on environmental issues. In the glorious words of John Muir: 'When we try to pick out anything by itself, we find it hitched to everything else in the universe.'

Holism is a strategy toward the goal of sustainability – human habitation of our planet without compromising environmental quality, without destroying the other components of the biosphere. If the goal is sustainability, then *everything* must be geared toward that goal. Our fervent hope for our book is that it will supplement your knowledge and skills as a tool helpful for your own journeying as an environmental communicator. Whether you are an environmentalist, an industrialist, part of a community group, or an individual trying to make a difference, we hope this book offers a way to meet your objectives.

Keep up the good work.

How We Organized This Book

More than anything, we have tried to do the winnowing for you, removing the less-than-essential to leave only the most valuable nuggets. We want our work to be readable and manageable in size, while offering high-value content. If you desire more information, we have included a 'Further reading' within the reference list at the end of each chapter.

The book is organized into three main sections. Part I sets the stage, giving a brief conceptual framework for the field of environmental communications. Part II delves into the basics of communication planning. Planning is to communicating what cartography is to travel. If you have a suitable and accurate map, getting there – wherever 'there' may be – is much, much more likely to happen. We want you to arrive where you expect. Randomly setting out on a journey without a clear direction can be exciting, but can take you to destinations you would be better to avoid.

Part III can be thought of as our toolbox, from which you may select any of many communication competencies. Chapters 10–15 cover skills and applications you need when working directly with people. Successfully interacting with people requires a whole box of tools, which come more naturally to some and not others. These tools range from speaking dynamically to an audience, to understanding why people think and act the way that they do. At Chapter 16, we broaden from interpersonal communicating to mass mediation of messages. The most prolific source of environmental information today is the news media, including those on the Internet. With help, this pervasive system can be navigated, to ensure messages are reported as intended. Chapters 17 and 18 review conflict management and risk. Conflict is a given in life, especially with human–environment interactions. Finally, Chapters 19 and 20 cap our discussions with an overview of communications as means of changing harmful behaviors and as a way of solidifying societies' reorientation to sustainability, our ultimate goal.

Reference

Safina C (2008) Let every tongue speak and each heart feel. In: Benjamin M (ed.) A passion for this Earth: writers, scientists, and activists explore our relationship with nature and the environment. Greystone and David Suzuki Foundation, Vancouver.

Acknowledgments

The impetus for writing this book was twofold. First, at a conference of environmental communicators in Chattanooga in 1995, a discussion arose about the need for a comprehensive and up-to-date introductory communications text for university undergraduate students and professionals in the field. Second, at The Ohio State University, a communications course is required for all students seeking a bachelor's degree from the School of Natural Resources. The packet of materials for that course evolved over many years, and transferred this myriad of handouts, readings, ideas, and exercises, plus a myriad collection of other ideas into a unified text that became the first edition, and now the second edition.

Although each chapter was written by the authors, it was in some way based on not just our ideas but on many other colleagues' ideas developed over several years. Their comments and feedback were truly invaluable. We specifically thank Dr. John Disinger, Dr. Rosanne W. Fortner, Dr. Joseph E. Heimlich, Dr. Gary W. Mullins, Dr. Robert E. Roth, and Dr. Stefan Somer for ideas and comments with preliminary versions of the chapters in the first edition. Dr. Nicholas J. Smith-Sebasto is gratefully acknowledged for his help with the first edition on the preliminary draft of the outline.

Although this second edition is our own expanded tome, thanks for helping hands and thoughts go out to Bridget Arend, Holly Hitzemann, Ken Shearer, Pamela Hill, Vivian Puxian, Jeff Brenman, Pawan Jaswal and Tammy Herring. Gratitude also goes to the unsung heroes working as research librarians at Denver Public Library, Harvard University's Museum of Comparative Zoology, The Ohio State University and Washington University Law Library.

We thank close friends and family members for their incredible patience and support.

Finally, to our editors over the long haul of this second edition, Judith Terpos and Tamara Welschot, our grateful thanks for their support, and patience.

The authors acknowledge that there are instances where they were unable to trace or contact the copyright holder for permission to reproduce selected material in this volume. The authors have included complete source references for all such material and takes full responsibility for these matters. If notified, the publisher will be pleased to rectify any errors or omissions at the earliest opportunity.

Richard R. Jurin, Donny Roush and Jeff Danter

Contents

Part I Principles of Environmental Communication

1 Understanding the World Around Us 3
    1.1 Introduction 3
    1.2 Axioms for Environmental Communications 3
    1.3 A Brief History of Environmental Communication 5
          1.3.1 Nature Writing 7
          1.3.2 Outdoor Recreation and Travel Writing 7
          1.3.3 Science Writing 7
          1.3.4 Public Affairs Reporting 8
          1.3.5 Persuasion 8
    1.4 The Growth of Environmental Communication 9
          1.4.1 1969–1974 12
          1.4.2 1989–1994 13
          1.4.3 2002 Onward 13
    1.5 Definitions of 'Environmental Communication' 13
    1.6 Models of Environmental Communication 15
          1.6.1 Communicating Environmental Information Model 15
          1.6.2 Ecological Model of the Communication Process 18
    1.7 A Sense of Place 19
    1.8 What Is Sustainability? 20
    1.9 Education for Sustainable Development 22
          1.9.1 Case Studies: Education for Sustainable Development (ESD) 23
    References and Further Reading 24

2 Communicating About the Environment 27
    2.1 Introduction 27
    2.2 Communication Modeling and Theory 28
          2.2.1 Communication Perspectives 30
          2.2.2 Message Elaboration 30
    2.3 What Are the Differences and Similarities Among Environmental Communication, Environmental Education and Environmental Interpretation? 31

	2.4	Principles of Adult/Community Education	33
	2.5	Implications for the Professional	34
		2.5.1 Case Study: Environmental Education Project WILD	35
		2.5.2 Case Study: Environmental Interpretation. Great Barrier Reef Marine Park	36
		2.5.3 Case Study: Environmental Communication	37
	References and Further Reading		37
3	**Developing Your Environmental Literacy**		**41**
	3.1	Introduction	41
	3.2	Literacy	42
	3.3	Numeracy	42
	3.4	Science Literacy	43
	3.5	Environmental Literacy	45
		3.5.1 Degrees of Environmental Literacy	45
		3.5.2 Measuring Environmental Literacy	46
	3.6	Ecological Literacy	48
		3.6.1 Nature-Deficit Disorder	49
		3.6.2 No Child Left Inside	49
	3.7	How Science Information Becomes Reliable	50
		3.7.1 Frontier Science	52
		3.7.2 Primary Literature	52
		3.7.3 Secondary Literature	52
		3.7.4 Textbook Science	53
		3.7.5 The Internet and the Knowledge Filter	53
	3.8	Thinking Critically About Scientific Information	54
		3.8.1 Does Their Argument Make Sense?	54
		3.8.2 Who Is the Source of the Information?	55
		3.8.3 Are the 'Facts' Placed in a Context of Accepted Knowledge?	55
		3.8.4 How Was the Information Obtained?	55
		3.8.5 What Kind of Study Was Reported?	56
		3.8.6 Were Measurements and Statistics Used Properly?	56
	3.9	The Art of Argumentation	57
	3.10	Case Study: Environmental Literacy. Last Child in the Woods	60
	References and Further Reading		61
4	**Investigating Environmental Issues**		**63**
	4.1	Introduction	63
	4.2	Components of Issue Analysis	64
	4.3	How Issues Arise	65
	4.4	Dissecting Issues	66
	4.5	Value Descriptors	68
	4.6	Global vs. Regional vs. Local Issues	69
	4.7	Framing and Framing Anew	70

	4.8	Case Study: Environmental Issue Analysis. Ohio Beverage Container Deposit Legislation	71
		References and Further Reading	72

Part II Communication Planning

5 Planning Environmental Communications 75
 5.1 Introduction 75
 5.2 A Process for Planning Campaigns 75
  5.2.1 Problem Statement 76
  5.2.2 Goals 77
  5.2.3 Audience Analysis 77
  5.2.4 Objectives 77
  5.2.5 Message Development/Media Options/Audience Suitability 78
  5.2.6 Media Choice and Design 79
  5.2.7 Timeline 79
  5.2.8 Front-End Evaluation 79
  5.2.9 Formative Evaluation 79
  5.2.10 Summative Evaluation 80
  5.2.11 Project Budget 80
 5.3 An Outline for Writing a Communication Plan 80
 5.4 Case Study: Environmental Communication Planning 81
 References and Further Reading 82

6 Analyzing Your Audience 83
 6.1 Introduction 83
 6.2 Internals Versus Externals 84
 6.3 Population Segmentation 85
  6.3.1 Adoptions of New Ideas 85
  6.3.2 Support of Pro-environmental Issues 86
  6.3.3 Fragmentation, Selectivity and Loyalty 87
 6.4 Adopting New Ideas 88
  6.4.1 Awareness 89
  6.4.2 Interest 89
  6.4.3 Evaluation 89
  6.4.4 Trial 90
  6.4.5 Adoption 90
 6.5 Beliefs, Values, Attitudes, Worldviews, and Opinions 90
  6.5.1 Belief 91
  6.5.2 Value 92
  6.5.3 Attitude 93
  6.5.4 Worldview 93
  6.5.5 Opinion 93
  6.5.6 Situational Factors 93
 6.6 Memes 94

	6.7	Locus of Control	95
	6.8	A Model of Citizen Participation	97
		6.8.1 Entry Level Variables + Ownership Variables + Empowerment Variables → Environmentally Responsible Behavior	97
	6.9	Motivation	98
		6.9.1 Motivational Needs Models	99
		6.9.2 How to Motivate Adults	100
	6.10	Consumerism as a Way to Understand Preferences	101
		6.10.1 Business Communication to Assist Consumer Choice	102
	6.11	Case Study: Audience Analysis. Environmental Radio Soap Opera for Rural Vietnam	103
	References and Further Reading		103

7 Evaluating Your Messages' Effects ... 107
  7.1 Introduction ... 107
  7.2 Purposes of Evaluation ... 107
  7.3 Methods of Evaluating ... 108
    7.3.1 Surveys ... 108
    7.3.2 Participant Observation ... 110
    7.3.3 Interviews ... 111
    7.3.4 Group Consensus ... 112
    7.3.5 Secondary Analysis/Case Study ... 113
    7.3.6 Professional Judgment/Expert Opinion ... 114
  7.4 Quantitative Versus Qualitative Techniques ... 114
  7.5 Types of Evaluation ... 115
    7.5.1 Formative ... 115
    7.5.2 Process ... 115
    7.5.3 Outcome ... 115
    7.5.4 Impact ... 116
  7.6 Factors Influencing Evaluations ... 116
    7.6.1 Cost ... 116
    7.6.2 Expertise ... 116
    7.6.3 Risk of Failure ... 117
    7.6.4 Sample Make-Up/Selection ... 117
    7.6.5 Utility ... 117
    7.6.6 Timeliness ... 117
    7.6.7 Autonomy ... 118
  7.7 Evaluation Plan ... 118
  7.8 Case Study: Evaluation. Global Education Project in Central Asia ... 119
  References and Further Reading ... 120

8 Characterizing the Mass Media ... 123
  8.1 Introduction ... 123
  8.2 Convergence ... 123

	8.3	Characteristics of Mass Media	125
		8.3.1 Purpose	125
		8.3.2 Providing Information	126
		8.3.3 Persuasion	127
		8.3.4 Entertainment	127
		8.3.5 Audience Focus and Depth	128
		8.3.6 Delivery Channel	129
		8.3.7 Timeliness	130
		8.3.8 Cost	131
	8.4	Conclusion	132
	8.5	Case Study: Converged Media. 'Earth Song' by Michael Jackson	132
	References and Further Reading		133
9	**Highlighting Useful Media**		**135**
	9.1	Introduction	135
	9.2	Traditional Media	136
		9.2.1 News Releases	136
		9.2.2 Letter Writing	137
		9.2.3 Abstracts and Executive Summaries	139
		9.2.4 Public Service Announcements	139
		9.2.5 Information Sheets	141
		9.2.6 Science Writing	142
		9.2.7 Direct Mail	142
		9.2.8 Newsletters	143
		9.2.9 Interpretive Talks/Presentations	143
		9.2.10 Films	144
	9.3	New Media	145
		9.3.1 The World Wide Web	145
		9.3.2 Email	146
		9.3.3 Mobile Device Messaging	147
		9.3.4 Blogs	147
		9.3.5 Social Networking	148
	9.4	Converged Media	148
		9.4.1 Reader Responses	149
		9.4.2 Podcast Tour Guides	150
		9.4.3 Viral Marketing	150
		9.4.4 Webinars	151
		9.4.5 Streaming Events	152
	9.5	Unusual Media for Environmental Communication	153
	9.6	Case Study: Useful Media. Plant a Billion Trees (http://www.plantabillion.org/)	153
	9.7	Case Study: Useful Media. Motorola Renew and Samsung Reclaim	154
	References and Further Reading		154

Part III Skills Building and Practical Applications

10 Grouping Together Well 159
    10.1 Introduction 159
    10.2 Why Do Groups Exist? 159
    10.3 Community Groups and Their Special Aspects 160
        10.3.1 Hegemony 160
        10.3.2 Empowerment 161
        10.3.3 Revelation 161
        10.3.4 Education as Intervention 161
        10.3.5 Leadership and Dependence 161
        10.3.6 Openness 162
    10.4 Team Building Techniques 162
        10.4.1 Group Climate 163
        10.4.2 Building Relationships 163
    10.5 Capacity-Building and Civic Agency 165
    10.6 Managing Versus Leading 166
    10.7 Some Management Theory Ideas 167
        10.7.1 Contingency Theory 167
        10.7.2 Chaos Theory 167
        10.7.3 Systems Theory 168
        10.7.4 Self-Directed Teams 169
    10.8 Emotional Intelligence 169
        10.8.1 Trust as the Cornerstone of Empathy 170
    10.9 Formats for Presenting Information to Groups 171
        10.9.1 A Speech, Film or Demonstration 171
        10.9.2 Brain-Storming 172
        10.9.3 Buzz Sub-Groups or Small Discuss Sub-groups ... 172
        10.9.4 Role Playing 172
        10.9.5 Panel Discussion 172
        10.9.6 Colloquy, or Talk-Show Format 173
        10.9.7 Symposium 173
    10.10 Conclusion 173
    10.11 Case Study: Environmental Group Formation. 173
        Taiwan's Environmental and Sustainability
        Non-governmental Organizations 173
    References and Further Reading 174

11 Differing Ways of Thinking and Doing 177
    11.1 Introduction 177
    11.2 Personality Types 177
        11.2.1 Satir Modes 178
        11.2.2 Myers–Briggs Personality Typing 179
        11.2.3 Enneagrams of Personality 180

		11.3 Learning and Coping Preferences	181
		11.3.1 Field Dependent vs. Field Independent	181
		11.3.2 Gardner's Multiple Intelligences	181
		11.3.3 Learning Styles	184
	11.4	Accommodating People with Disabilities	184
	11.5	Conclusion	185
	11.6	Case Study: Multiple Intelligences and Learning Styles	186
	References and Further Reading		186
12	**Communicating Across Cultures**		189
	12.1	Introduction	189
	12.2	Culture: Macro Versus Micro	191
	12.3	Cultural Adaptation Theories	192
	12.4	Worldviews	193
	12.5	Where Are We Headed?	196
	12.6	Empathy or Apathy?	196
	12.7	Stereotyping Versus Sociotyping	197
	12.8	Sensitivity for People with Disabilities	198
	12.9	Cultural Awareness/Sensitivity/Competency	199
	12.10	Becoming Culturally Competent	200
		12.10.1 The Platinum Rule	201
		12.10.2 Why Do We Have Cultural Conflicts?	201
	12.11	Conclusion	201
	12.12	Case Study: Communicating Across Cultures. Cultural Context at Work	202
	References and Further Reading		202
13	**Speaking to an Audience**		205
	13.1	Introduction	205
	13.2	Structuring the Presentation	205
		13.2.1 Interpretive Theming	205
		13.2.2 A Presentation's Introduction	207
		13.2.3 A Presentation's Main Body	207
		13.2.4 A Presentation's Conclusion	208
	13.3	Delivering the Presentation	208
		13.3.1 Verbal Delivery	208
		13.3.1.1 Vocal Qualities	209
		13.3.1.2 Mannerisms and Posture	211
		13.3.1.3 Influencing Audience Emotions	212
	13.4	Overcoming Anxiety About Public Speaking	213
	13.5	Case Study: Public Speaking About and for the Environment	218
		Speaking of Earth: Environmental Speeches that Moved the World	218
	References and Further Reading		218

14 Communicating Without Words 221
14.1 Introduction 221
14.2 Kinesics: Physical Movement 222
14.3 Proxemics: Personal Space 222
14.4 Semiotics: The Science of Symbols 223
14.5 Paralanguage 225
14.6 Psycholinguistics 226
14.7 Metaphors 227
14.8 Cultural Implications 227
14.9 Consistency in Using Nonverbals 228
14.10 Conclusion 229
14.11 Case Study: Nonverbal Communication.
Nonverbals between Superior and Subordinate Workers 229
14.12 Case Study: Proxemics. 'Personal Space'
functional artwork by Vivian Puxian 229
References and Further Reading 230

15 Using Visual Aids 231
15.1 Introduction 231
15.2 Visual Aid Basics 231
15.3 You Don't Always Have Electricity 233
15.4 Authentic Items and Models 233
15.5 Warm Fuzzies 234
15.6 Flipcharts, Chalkboards and Whiteboards 234
    15.6.1 Flipcharts 235
    15.6.2 Chalkboards 236
    15.6.3 Whiteboards 236
15.7 Handouts 236
15.8 Even When You Do Have Electricity 237
15.9 Overhead Projectors and Transparencies 237
15.10 Slides 238
15.11 Video and Audio Clips 239
15.12 Computer-Generated Images and Programs
(PowerPoint, Keynote) 239
15.13 Conclusion 241
15.14 Case Study: Using Visual Aids. 'Thirst' Presentation
by Jeff Brenman of Apollo Ideas 242
References and Further Reading 245

16 Dealing with the News Media 247
16.1 Introduction 247
16.2 What Is the News Process? 247
16.3 Role of the Media 248
16.4 News Reporting Constraints 248
16.5 Accuracy in News vs. Accuracy in Science 249
16.6 Other Limitations to Science and Environmental Reporting 250

Contents   xix

    16.7 News Releases.. 251
    16.8 News Media Options.. 252
    16.9 Scientists/Engineers and the News 253
    16.10 Conclusion 254
    16.11 Case Study: Media Relations.
            Environmental Working Group.. 255
   References and Further Reading.. 255

17 **Managing Conflict**.. 257
    17.1 Introduction... 257
    17.2 Values of the Environment.. 257
    17.3 Reasons for Conflict 258
    17.4 Anatomy of Conflict 258
    17.5 Resolving Disputes 260
    17.6 Communicating About Conflict....................................... 261
    17.7 Conflict Happens 262
    17.8 Conclusion 263
    17.9 Case Study: Conflict Management in the United States
            Allagash Wilderness Waterway, Maine 263
    17.10 Case Study: Conflict Management in Mongolia.
            Pastoralists vs. Miners.. 264
   References and Further Reading.. 265

18 **Communicating About Risk**... 267
    18.1 Introduction... 267
    18.2 What Is Hazard?.. 268
    18.3 Outrage.. 268
    18.4 Risk Acceptance.. 269
    18.5 Mass Media Reports 271
    18.6 Acknowledge Uncertainty to Communicate
            Risk Effectively... 271
    18.7 Final Thoughts 272
    18.8 Case Study: Risk Analysis. Apples and Alar................. 273
   References and Further Reading.. 274

19 **Learning from Marketing and Public Relations** 277
    19.1 Introduction... 277
    19.2 Marketing and Social Marketing.................................... 277
    19.3 Public Relations 279
    19.4 Propaganda.. 281
    19.5 Greenwashing 281
    19.6 Philanthropy as Communication..................................... 283
    19.7 Summary... 283
    19.8 Case Study: Marketing and Public Relations.
            Organic Food is Harmful?... 284

19.9 Case Study: Marketing and Public Relations. U.S. Environmental Protection Agency and 9/11 Pollution 284
References and Further Reading 285

20 Walking the Talk of Green Business and Sustainability 287
20.1 Introduction 287
20.2 Corporate Social Responsibility 288
20.3 Frameworks for Sustainable Business Practices 289
    20.3.1 Hannover Principles 289
    20.3.2 Sanborn Principles 290
    20.3.3 Principles of Ecological Design 291
    20.3.4 Leadership in Energy and Environmental Design 292
    20.3.5 A Sense of Place for Businesses 292
    20.3.6 Corporation 20/20 294
    20.3.7 Green to Gold 294
20.4 Thinking Differently, Thinking Systemically 295
20.5 Corporate Sustainability Reporting 296
20.6 World Business Council for Sustainable Development 296
20.7 The Fourth Quadrant and the Green Collar Economy 297
20.8 Case Study: Walking the Talk in the United States LEED Platinum Certification for the Leopold Center, Baraboo, Wisconsin 298
20.9 Case Study: Walking the Talk in Sweden Corporate Sustainability Reporting by Svenska Cellulosa Aktiebolaget SCA, Sweden 299
20.10 Epilogue 300
References and Further Reading 300

Index 303

Bio Blurbs

Richard Jurin is a Biological Education Associate Professor at the University of Northern Colorado and the Coordinator of the Environmental Studies Program. Prior to moving to Colorado he was a lecturer at The Ohio State University in the School of Natural Resources (OSU/SNR). He gained both his second masters in Environmental Communication, and a Ph.D. in Environmental Education, Communication, and Interpretation with an Adult/Community focus from OSU/SNR. In a previous life he was a research Biologist/Biochemist with a degree equivalent in Biology and a master's degree equivalent in Biochemistry (both from Huddersfield University, England). He has served on the boards of the North American Association for Environmental Education and chaired two of their Commissions (Research and Sustainability Education), and the Colorado Alliance for Environmental Education among his many service components. He was also on the board of the Air & Waste Management Association (AWMA), East Central Section as the Environmental Communication officer. He is active in the National Association for Interpretation, is certified as an Interpretive Planner, Trainer, and Guide, and an Officer in the College and University Academics section. Besides publishing in professional journals, he recently co-authored three children's books on Cougars in connection with The Cougar Fund. His research interests include worldviews as barriers to sustainability; interpretations of visual media perceptions during non-formal and informal educational/communication settings; perceptions of wilderness; sustainability and spirituality; business leadership for ecological sustainability; and sustainability in tourism and interpretation.

Donny Roush works as advancement director for The Odyssey School, a public charter school in Denver, Colorado, USA, specializing in experiential education. There he works on collaborations, grants, and service learning, while providing environmental education support to the school's faculty. He has worked for the Audubon Society of Greater Denver, the Idaho Museum of Natural History, Idaho National Laboratory, The Ohio State University, the Columbus Marathon, *Ohio Runner* magazine, and *Lake Superior Magazine*. From 2001–2006, he was the first-ever executive director of the Idaho Environmental Education Association (IdEEA), where his most proud accomplishment was building a network of 20 schools demonstrating environment-based education in the age of No Child Left Behind. IdEEA's accomplishments were recognized in 2006 when the organization was awarded Affiliate of the Year by the North American Association for Environmental Education. Degree-wise, he holds an MS in human dimensions of natural resources from The Ohio State University and a BS in magazine journalism from Bowling Green State University. Professionally, he serves as a certified facilitator for the Guidelines for Excellence in Environmental Education, National Environmental Education Advancement Project Leadership Clinic Model, and Project WET (Water Education for Teachers). In February 2009, he became one of the first six individuals in Colorado recognized as a Certified Master Environmental Educator.

Jeff Danter works as the Florida State Director of The Nature Conservancy. During his 10 years with the Conservancy, he has also served as the Alabama State Director and the Project Director of the Disney Wilderness Preserve. Prior to joining The Nature Conservancy, Jeff worked for The Ohio State University as an instructor and research associate where he engaged with several government clients. Jeff also served as a consultant to conservation agencies and non-profits. Prior to his career in conservation, he worked for 10 years for a large multinational corporation.

Jeff has three degrees from The Ohio State University (go Buckeyes!): a Ph.D. in Natural Resources Management, an M.B.A., and a B.S. in Chemical Engineering. His research centered on the institutional leadership of conservation and natural resource agencies engaged in ecological management. He has served on the School of Forestry and Wildlife Science Advisory Committee at Auburn University and currently serves on the Department of Wildlife Ecology and Conservation Advisory Council for the University of Florida.

Jeff and his family live in Orlando, Florida.

Part I
Principles of Environmental Communication

Chapter 1
Understanding the World Around Us

1.1 Introduction

We live on a unique and magnificent planet, a place of rare beauty and great value (Fortner 1991). And, every human on Earth crafts, exchanges, and receives messages about our home. We are all environmental communicators.

Each of us already partakes in the process that is subject of this book – environmental communication. If we are already doing it, then we can optimistically hope to learn to do it better. More clearly. More effectively. With wider and deeper meanings.

As communicating is a skills-based process, we can learn to improve our abilities to send and decode information-packed messages. Environmental communication finds its basis in an urgency to better understand and translate human relationships with the rest of nature. As every person attempts to create meaning from the sensations produced by the world around them, our population has grown during the last few centuries with rapidity usually found only in the microbial realm. We crave nature even as we crowd everything non-human.

Human beings have an innate affinity to work purposely through their relationships with the rest of the natural world (Cantrill 1999). More than anything, we seek meaning. We fervently want to understand the world around us, because we are bonded to it (Wilson 1984; Kellert 1993). As we move toward our own versions of individual clarity, we like to discuss it with others. Our discourse about the world around us is environmental communication.

1.2 Axioms for Environmental Communications

What do we know about environmental communication? Our discussion of environmental communication rests on a few foundational concepts. These principles provide footings for environmental communication's foundation.

Communication, as considered here, is a human activity – Though scientists have identified many processes between non-human organisms that can be labeled 'communication,' the deliberateness and richness of messages that form environmental communications is only found among humans. Rightly, the presumption of humans somehow being above the rest of nature has been fingered as a source of many of our environmental problems. It is both ironic and hopeful that this same distinction is now being used by environmental communicators to help overcome human-caused degradation of the biosphere.

You cannot not communicate – Mere existence is an act of communication; to be is to communicate. Trying not to send out any messages sends a message in itself. If one sends messages regardless, it would be wise to attempt to communicate with purpose and competence. For those choosing the natural resource professions, this is doubly true. Being understood depends on suitably forming and sending messages. Confusion is one likely result of poor communication.

Understanding is the goal of communication – Communication is successful when a message is comprehended by its intended recipient. Many messages go misunderstood, clouded by some glitch in the system. These glitches are called noise. Noise is to communication what entropy is to life, the thing against which the system constantly struggles. Noise can occur within any portion of a communication system. Communicators' strive to overcome and circumvent noise.

Most responsibility in this process rests with the communicator, not the recipient – Carefully and skillfully, successful communicators package their messages for maximum effectiveness. They should know exactly to whom they are sending specific information, how this audience prefers to receive such information, and how they can be expected to translate it. Knowing why one wants to send messages helps a communicator shoulder this responsibility successfully. So, ethics play a decisive role in environmental communication. When a message is not understood, the fault falls back on the originator of the message. The communicator then becomes the fixer.

Human society depends daily on nature for survival – Hold your breath and read this paragraph. Everything we do within our highly developed and specialized human society depends on the services provided to us by a living, healthy planet (Baskin 1997; Daily 1997). Earth is the only home we have and the functions of its biosphere sustain us. Natural systems give us clean air, clean water, food, shelter, pleasure, beauty and belief in affairs beyond ourselves. Further, our economy – the human institution most important in political deliberations around the world – depends on nature's economy. In short, we live and work if, and only if, nature lives and works. Exhale and celebrate your reconnection with the life-giving earth (Cohen 2007).

Earth has its own messages to share with us – Listening to the planet is one way of conceptualizing the work of science. Scientists are a crucial source of information for all environmental communicators. Environmental communicators' intermediate position between scientists and the larger non-technical population is precarious, exciting, and crucial. Those practicing environmental communication need to hone their own perceptual skills, to understand what the planet has to say through them. Human senses have been extended by all sorts of gadgetry and instrumentation.

We would understand much less about the state of Earth and the life it sustains without this technology. But, many commentators contend that our technology has blocked vital messages coming from the planet itself. Technology, the stuff of modern life, becomes noise in this interpretation. If we wish to learn to mute these particular distractions, environmental communication can help. Environmental communicators can, too.

Atop the sturdy foundation provided by these axioms, a conceptual framework is being built. Onto this framework we can attach a myriad of skills useful to environmental communicators. The principles and skills of environmental communication are examined in this book. We hope you find practical suggestions here and can apply them immediately in your work, for our world.

1.3 A Brief History of Environmental Communication

As long as humans have interacted with each other and with nature, there has been environmental communication. But, 'environmental communication' as a label, applied with wide agreement by practitioners and academics, has a much shorter history, dating back to 1969. In the vast expanses of academic literature, we attempted to pinpoint the first use of 'environmental communication.'

When *Journal of Environmental Education* debuted 4 decades ago, environmental education and communication were seen as conjoined, two sides of the same thing. In the first article of the first issue, titled 'What's New about Environmental Education?,' Schoenfeld (1969) defined environmental education as 'communication aimed at producing a citizenry that is knowledgeable concerning our environment and its associated problems, aware of how to help solve those problems, and motivated to work toward their solution.' Over the next decade, the 'it' was often labeled 'environmental education/communication.' There seemed to be more concern in distinguishing 'environmental' from antecedent descriptors such as 'conservation,' 'outdoor,' and 'nature,' than in splitting 'education' from 'communication.' Resolution tended toward a notion of encompassing all, a bias for interconnectedness, and a tendency to fixate on negative human impacts.

From surveys of our personal collections, home institution libraries, and on-line searches, we offer a selection of touchstones for environmental communication's emergence:

- 1949 – *A Sand County Almanac*, by Aldo Leopold, first published. Paperback edition in 1966 reaches millions.
- 1957 – *Interpreting Our Heritage*, by Freeman Tilden, first published.
- 1958 – *Environment*, first issue published by Scientist's Institute for Public Information.
- 1962 – *Silent Spring*, by Rachel Carson, first published.
- 1968 – Pictures of Earth from space, taken nearly as an afterthought by Apollo 8 astronauts, stun viewers who see a small, finite planet against a void. 'Spaceship Earth' metaphor is born.

- 1969 – Schoenfeld uses term 'environmental communication' in inaugural *Journal of Environmental Education*. He was a wildlife manager and newspaper reporter, turned university professor. The journal continues today.
- 1970 – First Earth Day follows the signing of the National Environmental Policy Act by 4 months. Both are visionary.
- 1971 – *Natural Resources and Public Relations*, by Douglas Gilbert, published.
- 1972 – United Nations Environment Programme founded, with main offices in Nairobi, Kenya.
- 1973 – 'Mass media and man's relationship to his environment,' by Gerhart Wiebe, appears in *Journalism Quarterly*.
- 1981 – 'John Muir, Yosemite, and the sublime response: A study of the rhetoric of preservation,' by Christine Oravec in *Quarterly Journal of Speech*, brings rhetorical analysis to bear on environmental communications. Up to this juncture, natural resource management and media studies served as predominant sources of scholarship.
- 1988 – *Reporting on the Environment: A Handbook for Journalists*, by Sharon and Kenneth Friedman, published.
- 1989 – Environmental Media Association founded, using entertainment celebrities to promote environmental awareness.
- 1990 – Society of Environmental Journalists founded, by North American reporters.
- 1991 – First scholarly Conference on Communication and Our Environment convened (though its inaugural name is Conference on the Discourse of Environmental Advocacy). Founded by Oravec, the conference continues biennially.
- 1993 – International Federation of Environmental Journalists founded, by European reporters.
- 1994 – *Electronic Green Journal* launched, one of the earliest on-line, open-access journals.
- 1995 – *Environmental News Network* launched, as an on-line news service.
- 2001 – *Applied Environmental Education and Communication* debuts.
- 2004 – *The Environmental Communication Yearbook* appears, lasting three volumes.
- 2007 – *Environmental Communication: A Journal of Nature and Culture* debuts, succeeding the yearbook format with a thrice-annual publishing schedule.

What preceded Schoenfeld's coinage? The roots of American environmental communications, reflected mostly in published artifacts and federal policies, go at least as far back as the late 1800s. Pioneers in examining the American interaction with the land and waters of the country have names you've probably heard before: Henry David Thoreau, John Muir, Fredrick Law Olmstead, George Perkins Marsh, John Wesley Powell, Gifford Pinchot, Theodore Roosevelt, Stephen Mather, Aldo Leopold, and Rachel Carson. This progression of writers and politicians encouraged Americans to think in environmental ways. Their writings are bulwarks of environmental history in the United States and have been influential worldwide.

While their perspectives did not always agree, they share a similarity – their subject was our environment in its totality. Throughout, a premise was pronounced: natural resources must be carefully conserved if we are to continue to thrive. In the early years, there was more of an emphasis on conservation of natural areas with prudent use of natural resources, rather than top-down regulatory control. Leopold gets most credit for emphasizing ecological thinking and holistic consideration of actions within our environment.

Looking at these antecedents of today's environmental communication magnifies our appreciation of today's situation. Schoenfeld (1981) asks, 'Irrespective of their roots, are there common denominators among the various forms of environmental communication? Yes. All are focused on a comprehensive rather than a compartmentalized approach to the people-resources-technology system. A basic theme in environmental communication hence is interdependence – that everything is connected to everything else.' He lists five roots from which environmental communications grow, selecting magazines as examples.

1.3.1 Nature Writing

During the Age of Discovery, explorers wrote of newfound (-to-them) lands, extreme weather events, and previously undescribed creatures. Early American writers like Muir and Thoreau wrote extensively of nature's acts. Nature writing continues to be found in the magazines of many environmental organizations, often under the heading of adventure writing. Magazines like *Field and Stream*, *Audubon*, and *National Geographic* are filled with tales of challenges in the natural world and phenomenology noticed through careful observation. TV documentaries, such as BBC's 2006 'Planet Earth' series, deal with nature with the same overwhelming sense of awe and detachment.

1.3.2 Outdoor Recreation and Travel Writing

Closely aligned with nature writing, many periodicals cover nature, outdoor recreation, and travel writing in the same issue. Early travel writers are credited with spurring mass emigrations to the United States. The journals and writings of the early pioneers who trekked across the Great Plains and Rocky Mountains toward new lands riveted readers on America's East Coast, in Europe, and elsewhere. Surviving examples of this genre are *Outside* and *Travel + Leisure*.

1.3.3 Science Writing

Many of the premiere science journals and magazines began publishing in the mid-1800s. *Scientific American* first appeared in 1845; *Nature* 1869; and, *Science*, in 1883. All three continue to publish and prosper. The reporting of science is not without controversy, as science has grown more specialized and esoteric. Still, a vision of wide public understanding of science remains, as expressed in *Nature*'s original mission statement: 'to place before the general public the grand results of Scientific Work and Scientific Discovery; and to urge the claims of Science to a more general recognition in Education and in Daily Life' (Nature 1869). Environmental communicators often face obstacles in producing clear and concise explanations of scientific findings. Almost a century ago, Slosson (1922) noted 'The would-be popularizer [of science] is always confronted by the dilemma of comprehensible inaccuracy or incomprehensible accuracy, and the fun of his work lies mainly in the solution of that problem.'

1.3.4 Public Affairs Reporting

Newspaper reports of government and business activities, the backbone of print journalism, dates to the 'muckraking' of the late 1800s and early 1900s, when newspapers found means to wrest themselves from the clutches of nineteenth century political parties and pursue independent investigations of threats to public health and safety. Such investigative journalism means to reveal graft, protect the public good, and force social change. Reporters advocate for the 'little guy' struggling against Goliaths of industry and other oppressors. They see themselves as giving voice to the voiceless. *Time* and *Newsweek* are the periodical exemplars of public affairs reporting. The Internet and more urbanized populations have morphed many news media into new media, as they adapt so as to remain a powerful force in shaping society and politics. The last 15 years has seen sea changes in the agenda-setting and gatekeeper roles formally consigned by large newspapers and network television newscasts. Access has become more instantaneous and audiences more fragmented. But, people still tend to use journalists and their products to find out what is happening in the world beyond their own senses.

1.3.5 Persuasion

Ancient Greeks – most famously, Aristotle – were the first masters of civic speaking and civil argumentation. Skilled application of the art of rhetoric equaled finding the right way to be persuasive in a particular situation. Persuasive writing, as opposed to oratory, can be traced back to the Middle Ages. Consider Martin Luther who penned, then nailed his challenges to doctrine to a church door. Within the last century, Gifford Pinchot, as head of the new U.S. Forest Service used persuasion to get Congress to set aside vast areas of the United States as national forests. National Parks, designated Wilderness areas, official Wildlife Refuges, and large National Monuments are evidence of his success, since most were carved out of the National Forest system. More recently, David Brower created several non-governmental organizations that collectively were able to persuade Congress to enact many laws to protect our environment. These groups and many of their offshoots still actively lobby Congress, publish magazines, and manage programs to help various portions of the environment. Other groups in opposition have also set up persuasive campaigns to counteract these forces. The greatest persuasive power rests with those appealing to current collective consciousness of an active portion of the populace and the prevailing zeitgeist. Magazines from leaders in this area include *Sierra* and *Earth Island Journal*, both of which have Brower to thank for their existence.

1.4 The Growth of Environmental Communication

With confidence, we can say there is an exponentially greater amount of environmental communication available now than when this particular type of information-exchange was first identified 4 decades ago. Conventional wisdom says the world of the early twenty-first century is awash in more information than could have been imagined around 1970.

Considering 'data' to be observations about the world recorded in some manner and 'information' to be data put into some sort of context, how much more new information is there in 2010 than existed in 1970?

About 1,000,000 times as much (Pool 1983; Lesk 1997; Lyman and Varian 2000, 2003).

Did you get that? One million times more information will be created this year, compared to 1970!

Measuring total global information flow uses uncommon prefixes attached to –byte, as in the familiar Megabyte. In the pre-Internet era of 1970, Pool (1983) indexed 0.5 Terabytes of new information available through all print and broadcast news media, plus all mail, telephone, telegraphs, facsimiles, and data communication. He noted that he was unable to measure most internal corporate communication and person-to-person conversation, but suggested those would not more than double the total. By 1977, he indexed 0.8 Terabytes of new information. Even then, he lamented 'information overload.' (A Terabyte equals 1,000 Gigabytes or 1,000,000 Megabytes.)

Lesk (1997) compiled a set of estimates of total information and new information creation for 1989. Overall, he estimated there to be 12 Exabytes of information in the world, with just over 1 Exabyte being created annually. His estimates may be high because new data compression technology was not considered, especially in considering telephone conversations. (An Exabyte equals 1,000,000 Terabytes.)

Come the 1990s, Internet use was doubling each year. 'Overload' gave way to a Gigabyte explosion. The 'How Much Information?' project at the University of California-Berkeley calculated 2 to 3 new Exabytes created in 1999, and a staggering 5 Exabytes of never before existing information generated and stored in 2002 (Lyman and Varian 2000, 2003). Consider electronic information flowing through the Internet and telephones – now mostly of the wireless variety, without cords – and we added another 18 Exabytes. Your per capita share of all this info: at least 800 Megabytes, with a lot more if you talk on the phone more than a few hours per month.

Information flows were clunky, linear, and controlled by relatively few gatekeepers on the first Earth Day, April 22, 1970. Those celebrating our home planet that day received most of their environmental information from printed and broadcast sources of news – TV, radio, newspapers, magazines, and books – with perhaps a few pamphlets, motion pictures, and phone calls thrown in. This was a time when the few computers there were filled entire rooms at either military installations or research universities. Telephones had cords, rotary dials, and were black and boxy.

The advent of the digital age means electronics near-at-hand for most people in developed countries. It means those in developing countries are catching up with levels of first-world TV and radio use. It means almost 1 billion Internet users, with the highest concentrations in Asia, Europe, and North America, trading more than 62 billion email messages daily (Lyman and Varian 2003). It means as much TV and other video content each day as was produced the entire year of 1970. Supply now has no technological limit and has grown exponentially.

But, a person still has only 24 h a day in which to consume information. Time remains a constant and a constraint. Even with a bit of linear growth in demand – from an estimated 700 to 1,000 min/day of media use per household – supply-to-demand ratio in 2005 was calculated at 20,934:1 (Neuman et al. 2009). Message-makers face stiffer and stiffer competition, even as they crank out more and more content. Message-readers become pickier with their precious attention, more adept at grazing multiple channels, and better able to mash together media and meanings. Consumers can seem more easily distracted and, simultaneously, better suited to pull the exact information they want and need from disparate, sometimes overlapping sources. For example, imagine a student in Australia wants to know about a concert by her favorite band playing that day in Iceland. Using her laptop, running an Internet browser, she performs a search to locate a music reviewer's web-log that's part of an on-line magazine posted to a social network. She instantly links to the writer's entry during which one of the band's better numbers plays. After reading and listening, she follows another link to a Canadian TV entertainment report about the band archived from the week before. The report plays using a high-definition media-viewer program already installed on her laptop. She likes the clip and emails it to a friend in Taiwan. All this takes less than 10 min to transpire; it covers the globe and crosses several media. Has she only been an Internet user, or was she also a magazine reader, recorded music listener, TV watcher, and personal letter writer?

Distinctions between media are blurring; cross-channel information flows are emerging as the new norm. People are more and more likely to 'pull' specific information they seek, rather than have it 'pushed' at them (Neuman et al. 2009). Wayfinding within the Internet realm becomes crucial to being informed rather than overloaded. Search engines, services which attempt to catalog Web content, play a keystone role. Half of U.S. Internet users used search engines in 2008, with higher

1.4 The Growth of Environmental Communication

frequency for college graduates, the wealthy, men, those under 30, and those with high-speed home connections (Fallows 2008). The Google search engine leads the way, with the most searches and the most sophisticated, if secretive, technology (Witten et al. 2007).

To borrow a term from biologists, we may be living in a period of 'punctuated evolution' in our communication systems. Rapid expansion in the amount of information has been matched by speedy shifts in the ways and means of moving words – both written and spoken – and pictures from sender to receiver. Twenty years ago, few had heard of the Internet. Fifteen years ago, search engines were more a dream than a reality. Ten years ago, no one had yet made a friend on Facebook or MySpace, or had an avatar within the virtual world of Second Life. Five years ago, no one had yet viewed YouTube and there was no such thing as an iPhone, a consumer electronic device combining phone, Web browser, recorded music player, camera, and data-storing personal computer (Fig. 1.1).

In short, more and more of the world's swelling information flow is moved electronically, the shelf-life of information has shortened dramatically on-line, and technology-supported decisions about individual information consumption have accelerated to match the rapid-fire rate at which new messages come at us.

Within the larger and ever-larger torrent of information, what can be said about the flow of environmental communication? Indications are environmental communication's expansion has outpaced the whole. More of our world's information is environmental in nature and more environmental information is being consumed. Empirical clues of the growth are 'the irruption in magazine environmental content' reflecting public abhorrence toward rampant industrial pollution in the

Fig. 1.1 The evolution of communication (cartoon). Credit: Mike Keefe, The Denver Post & InToon.com

Fig. 1.2 Environmental periodicals. Original graphic, Donny Roush

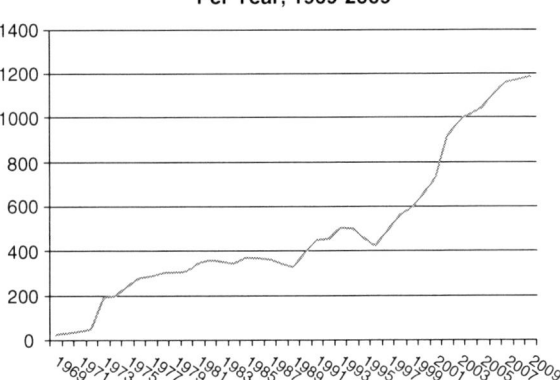

late 1960s and early 1970s (Schoenfeld 1982, 1983), scholarly production of environmental communication research at ten times the base-rate for the social sciences (Pleasant et al. 2001), and steep-sloped tallies of environmental-themed periodicals (Roush 1998).

We conducted a census of environmental periodicals from 1969 to 2009, using their numbers as proxy for overall volume of environmental communications (see Fig. 1.2). We base our use of this measure on Mott's (1939) theory of periodicals as accurate contemporary histories and one of the first repositories of evidence of cultural importance placed on social issues. Broadly, three undulations show on the 40-year arc of environmental communication's development. Wave crests show, though they take place amid a more apparent backdrop of steadily rising environmental communication action and are followed by no more than 3 years of contraction. Notable upticks in environmental communication activity occurred:

1.4.1 1969–1974

A mythical era of social change, the late 1960s and early 1970s were a time of fervent protest of just about all aspects of the socio-political status quo, where millions of mostly young people protested for the ideals of love, peace, and social harmony. Anti-pollution demands, while not as significant as anti-war messages within the ethos of the time, lead to improvements in environmental management and national legal frameworks within which industries functioned. The first wave of environmental periodicals swelled to 209 in 1974, from an initial 31 being published in 1969, as found for cataloging by *Environmental Periodicals Bibliography* when it launched in 1972.

1.4.2 1989–1994

Anniversaries can be motivating for mass movements. The 20th anniversary of Earth Day provided grist for a reemergence of environmentalism in the United States. Globally, the Rio Summit in 1992 reinvigorated attentiveness to environmental degradation, its symptoms, causes, and solutions, with human population and issues of social justice coming to the forefront. Our census of environmental publications shows a plateau across the 1980s, bumping to more than 500 for the first time in 1993.

1.4.3 2002 Onward

Environmental communications' third wave, the biggest to date, was rising by 2002 and may not have crested yet. More environmental publications carry more content than ever before. Now called *Environment Index*, the database we're using as a gauge now catalogs 1,187 periodicals. So, though fighting against a millionfold increase in 'info' might seem insurmountable, environmental messages are getting through. Environmental communicators are having success, as they ride the third wave.

1.5 Definitions of 'Environmental Communication'

'Environmental communication' has its own Wikipedia entry, as sure a sign of its arrival in the Internet Age as any amount of periodical publishing. (Wikipedia is a free, online encyclopedia written by everyone who cares to provide their input. Any information in Wikipedia can be peer-reviewed by all who read it and so is constantly vetted.) Here is the entry (as retrieved on August 3, 2008):

> Environmental communication refers to the study and practice of how individuals, institutions, societies, and cultures craft, distribute, receive, understand, and use messages about the environment and human interactions with the environment. This includes a wide range of possible interactions, from interpersonal communication to virtual communities, participatory decision making, and environmental media coverage. Environmental communication as an academic field emerged from interdisciplinary work involving communication, environmental studies, environmental science, risk analysis and management, sociology, and political economy. (Wikipedia 2008)

'Environment' and 'communication' are large, multifaceted terms, of the sort cultural scholars call 'constructs' because they are built from multiple connotations. As large linguistic symbols, constructs can pack a punch. Some communicative uses of constructs stimulate visceral and powerful reactions. Weighty constructs in post-modern Western culture include 'freedom,' the ever-popular 'love,' and the fast-rising 'terrorism.' Cultures rely on linguistic symbols to encapsulate and transfer values, beliefs, rituals, and behavioral norms. In short, we are dealing with a pair of constructs, melded together conceptually and considered as field of study and endeavor only in the last 4 decades.

A problem with defining a term made from a coupling of heavily-laden words is the difficult-to-escape need to be tautological. Few explanations of the meaning of 'environmental communication' manage to not use at least one of the two self-referentially. Though the editor of *Environmental Communication: A Journal of Nature and Culture* warns against 'succumbing to the trap of definition' (Depoe 2007), we wish to review what others, beginning with Schoenfeld and Wikipedia's anonymous contributors, have said about the meaning of 'environmental communication.' Then, we attempt to distill a short and accessible version.

Cox (2006), in his exhaustive and studious *Environmental Communication and the Public Sphere*, offers both an informal and a formal definition:

- Informal – 'a study of the ways in which we communicate about the environment, the effects of this communication on our perceptions of both the environment and ourselves, and therefore on our relationship with the natural world.'
- Formal – 'the pragmatic and constitutive vehicle for our understanding of the environment as well as our relationships to the natural world; it is the symbolic medium that we use in constructing environmental problems and negotiating society's different responses to them.'

Cox lists practical outcomes environmental communications might have: education, attention, persuasion, mobilization, and assistance. More subtly, he notes, environmental communication shapes our perceptions of the natural world as well as our interactions with and impacts on it.

Before referring readers to Cox's formal definition, The Environmental Communication Network – an on-line community of self-identified environmental communicators facilitated by the State University of New York College of Environmental Science and Forestry – states, 'environmental communication is all of the diverse forms of interpersonal, group, public, organizational, and mass communication that make up the social discussion/debate about environmental issues and problems, and our relationship to non-human nature.' (Meisner 2008)

Corbett (2006) pushes for an enlarged conception of environmental communication in her approachable yet cosmopolitan *Communicating Nature: How We Create and Understand Environmental Messages*. She posits 'environmental communication is:

- Expressed in values, words, actions, and everyday practices
- Individually interpreted and negotiated
- Historically and culturally rooted
- Ideologically derived and driven
- Embedded in a dominant societal paradigm that assigns instrumental value to the environment and believes it exists to serve humans
- Intricately tied to pop culture, particularly advertising and entertainment
- Framed and reported by the media in a way that generally supports the status quo
- Mediated and influenced by social institutions like government and business'

On the next-to-last point in Corbett's list, Cox (2007) agrees, articulating the position that despite normative inclinations, environmental communications has an ethical duty. Like conservation biology, which aims to save imperiled species and ecosystems, and oncology research, which seeks to eradicate cancer, environmental communication is a 'crisis discipline' tasked with creating ways to convince people to reverse degradation of the planet. The purpose of the field, he writes, is 'to enhance the ability of society to respond appropriately to environmental signals relevant to the well-being of both human communities and natural biological systems.'

Two other thinkers offer aligned ideas. In the first two editorials for *Applied Environmental Education and Communication*, Day portrays environmental communication as 'new' (2001) and then as a 'weapon' (2002), to be wielded against dirty water and other environmental health hazards which annually kill millions. In the first *Environmental Communication Yearbook*, Senecah et al. (2004) note the evolution of a distinctive tradition, or 'canon,' for environmental communication, which supports authoritative voices capable of influencing business, health and natural resource policy.

With due respect, we offer our own definition, based on those above and this chapter's conceptual framework:

Environmental Communication – the systematic generation and exchange of humans' messages in, from, for, and about the world around us and our interactions with it.

1.6 Models of Environmental Communication

Like definitions, models are a staple of introductory chapters. While depicting the complex graphically is a reductionist exercise, it does allow us to picture an entire system, with hopes of better understanding its elements, structures, inputs, outputs, and flows. Environmental communication has, so far in our text, been noted to be a 'process,' 'study,' 'practice,' 'vehicle,' and 'system.' Elsewhere, others have labeled it an 'activity/phenomenon' (Meisner 2008) and a 'nexus' (Day 2001; Depoe 2007). All terms apply; none is inaccurate. Still, what we have so far is a set of valid labels and a series of thoughtful stabs at a definition. Might a visual help to further our understanding?

To that end, we offer two models: the Communicating Environmental Information Model (expanded and adapted from Witt 1973) and the Ecological Model of the Communication Process (adapted with one addition from Foulger 2004).

1.6.1 Communicating Environmental Information Model

With its many arrows, this model helps to illustrate the multitude of inflows, outflows, and cross-flows of information from various actors, as well as how types of acknowledged message-makers are embedded in the larger public. Arrows represent

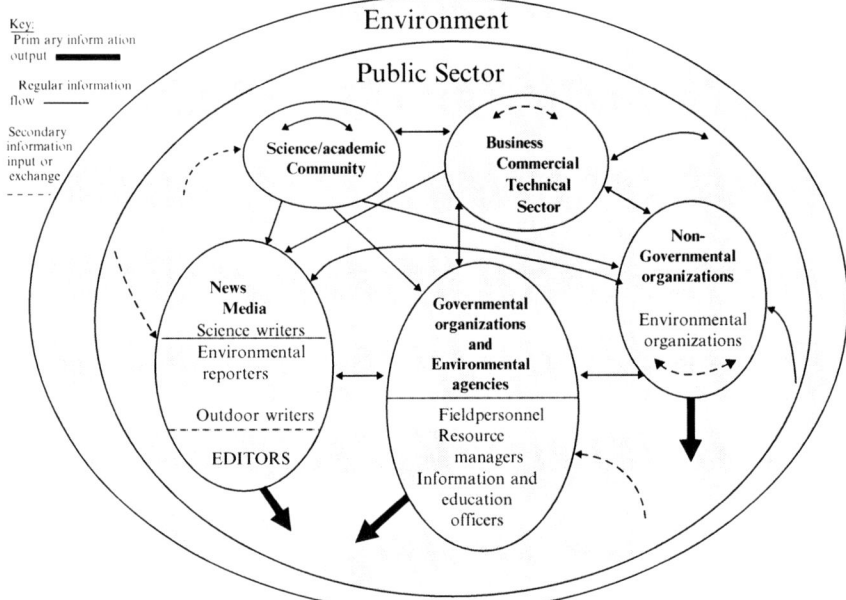

Fig. 1.3 Communicating environmental information model

major information flows, with usual strength of flow matching weight of the line. The model suggests the environment as stage, backdrop, inspiration, and milieu, or perhaps the ether through which communications flow.

As the Communicating Environmental Information Model was first devised well before the advent of the Internet, it shows groups of people as the sources and receivers of messages, still within our singular environment. The Internet may have drastically increased the volume and speed of messages, but this bigger, faster and wider pipeline for information has not changed the system's optimal need for human creation and consumption of messages (Fig. 1.3).

Consider some generalities about the interacting groups:

- **Scientific/Academic Community** – This sector tends to be the source of information which has undergone the most vetting and scrutiny. Many societal decisions, especially governmental policy, rely on scientific information for their rationale. A millionfold increase in information flow leaves little time for peer-review of most data contextualized into some sort of information. Hence, reliability issues have risen. Nevertheless, scientific journals are still the preeminent channel for the most respected communications. But, these journals' audiences are, by and large, other scientists. Transfers to the public sector require translations often as substantial as from one language to another.

  - Scientific/Academic Community to Business/Commercial/Technical Sector often involves collaborative research, some of which may be secretive, all of

which tends to be expensive. Many academic labs are home to the necessary expertise, yet lack adequate funding. Businesses provide the financial capital in exchange for confidentiality. The public sector may not receive the new information for some time.
- Scientific/Academic Community to News Media and/or Governmental organization and Environmental agencies involve review and application of findings by the science community. Alternatively, these two sectors may seek out experts in order to guide decision-making processes.
- Non-governmental organization sector tends to receive information filtered through News Media.
- Public Sector has some diffuse power to affect further funding of science, through communication – whether supportive and protestive – with agencies such as the U.S.'s National Science Foundation and the European Research Foundation. This group represents the most common source for raw information for environmental communicators,to repackage into more refined message-products.

- **Business/Commercial/Technical Sector** – Competition encourages certain communications, through attempts to distinguish one's own goods and services from the rest. What we're talking about here is hawking one's wares – via advertising and marketing. As organizational success is linked to the bottom line, there is less cooperation between groups, so much information is deemed proprietary. Primary external audiences are the public in their role as customers. Purchasing is a form of feedback. To a lesser extent, the sector also communicates with governments to fulfill legal requirements.
- **News Media** – Their primary role is reporting and analysis of current affairs, often prompted by competitive spirit (i.e., not to be 'scooped'). The primary audience is the public sector. Secondary interactions will be with their sources within governments, the scientific community, business, and non-governmental organizations, roughly in that order of frequency. More than the other groups, the news media have been changing, trying to adapt to the new communications reality.
- **Governmental Organizations and Environmental Agencies** – Policies (in the form of laws, rules and regulations) require compliance and enforcement action between governments and the business/commercial/technical sectors. Some of their exchanges are available as public record, some is not. The public sector is also a primary audience for outreach and education efforts, through mass media as well as at governmental park units and public lands. The public sector has a smaller feedback role, by alerting agencies and officials to issues of concern.
- **Non-governmental Organizations** – NGOs sustain a whole persuasive communication network of their own, using environmental information. Primary audiences are the public sector through member communications, mass media, and pressure on government representatives. While many groups within this sector interact regularly, others tend to be isolationist. Most are mission-driven to focus on issue- or geographically-defined interests. Secondary audiences will be

business sector partners who can provide funding and policy-change support, when in their corporate interest. These communicative relationships can show strains when an NGO seeks to modify business behavior. NGOs reach out to news media as a means of expanding situation, event and organizational awareness. With NGOs dependent on public support, one can observe constant, often personalized messaging via direct mail and email, seeking affiliation through memberships and donations. Fund-raising may appear to be a primary function of many NGO groups because of the solicitous content. Examples of these kinds of groups are World Wildlife Fund, Greenpeace, and the International Union for Conservation of Nature.

1.6.2 Ecological Model of the Communication Process

Foulger (2004) put forth a new model of communication in an effort to update and refine basic theoretical material little changed in more than half a century. We like his representation, in large part, because it recognizes similarities in ways to show interactions between organisms in natural ecosystems and the human-made world of information. He gamely named his work the Ecological Model of the Communication Process (Fig. 1.4).

To Foulger, communication involves creator-consumers and their messages. Moving messages between minds requires language and media. Relationships among creators and consumers are dynamic, cyclical, and multifaceted as meanings are molded, traded, and reacted to. Creation and consumption of messages happen in tandem, often simultaneously, within individuals.

We extend Foulger's provocative representation by adding the environment behind all the other elements, as a place within which creator-consumers operate

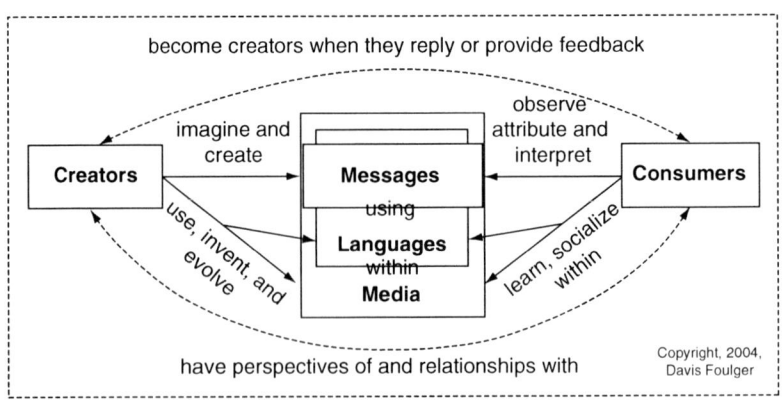

Fig. 1.4 Ecological model of the communication process (Foulger 2004)

and a stage where the messages are made and traded. The whole picture created looks like a depiction of energy flows in an ecosystem. An allusion to predator–prey cycles should not be missed. Similarities between models of ecological interactions and information flows have been noted by many (e.g., Schoenfeld 1981). Foulger's refinement creatively extends the idea of information as ecology.

A key to understanding the actions shown in the Ecological Model is the concept of 'instantiation,' the making of something concrete from something abstract. Meaning starts inside a person's mind. It is intangible, until a person creates a message using a language and a medium. As a message-maker, the person has instantiated ideas. Meaning has been made tangible. Foulger (2004) further notes that languages and media evolve over time, and are therefore also part of the creation of communication. Using language and media involves skill. He notes, 'People must learn language and media in order to be able to create and interpret messages. People need to know how to use language and media in order to communicate.'

You may have noticed the Communicating Environmental Information Model deals with group exchanges, whereas the Ecological Model of the Communication Process focuses person-to-person. The first model better captures the big picture, 'macro-communication' if you will; the latter, brings the process down to individual levels, or maybe 'micro-communication.' Historically, this division was noted as between mass communication and interpersonal communication. As communication has moved on-line and exploded in volume, such a distinction has blurred and blended. Nonetheless, we consider all communication as a human process arising from where we live. The place of communicating is our environment.

1.7 A Sense of Place

Much of the content of environmental communication tells about the places in which people live and function. Effective messages connect and relate to what people perceive as their particular place. A person may not live physically in a 'place' they care about. They can still have connections to places outside of where they live. Below we discuss sustainability. To achieve sustainability, we all must have a sense of the planet as a place.

Place is defined in both environmental (geophysical and location) and social (community and culture) dimensions. A sense of place organizes around meanings individuals and groups give to a location and the specific qualities of that setting. The meaning deepens through events that occur as part of everyday life and collective experience within a community.

There are three basic dimensions of place (Steele 1981):

Psychological (Place Attachment)

- Place dependence – use of area for professional or recreational purposes
- Place identity – understanding and conception of self within a particular setting; personal history, and anticipated future, within that setting

Social (Community and Culture)

- Inscription of 'sense of place' through cultural processes
- Social networks within place
- Familial ties to place
- Political and environmental involvement/activism

Political/Economic

- Local, regional, state, national, political boundaries and norms
- Opportunities for collaborative action focused on place-based interests and needs
- Political and environmental involvement/activism

1.8 What Is Sustainability?

One of the most widely used definitions of 'sustainable development' comes from Brundtland (1987): 'Sustainable development is development that meets the needs of the present without compromising the ability of future generations to meet their own needs.' While introducing the idea of equity, the Brundtland definition does not express the full range of ideas necessary to achieve sustainability. We have decided to use a newer version here, 'The ability of the biosphere to perpetually renew itself, yet still allow humans the ability to derive resources to live prosperously and in harmony with nature indefinitely.'

Obviously, the term prosperity is a crucial point of contention. Many people would happily say they want to be a millionaire, but if given a million dollars and simultaneously set on a small desert island in the middle of the Pacific Ocean would not feel prosperous. If we define prosperity as enhanced well-being rather than increased Gross Domestic Product (GDP), communicating concepts such as sustainability and issues about the environment take on different perspectives. 'Sustainability' is a continually evolving term and so fraught with a need for extensive discussion with each modification. Our role as communicators is to be aware of different perspectives and to help clarify them for our audiences.

In the United States, sustainable development is generally considered to have the three intertwined components: economy, society, and environment. Together, these comprise the Triple Bottom Line (see Fig. 1.5). The left side shows how a less-than-sustainable society treats the components as separate parts, while the right side indicates that all three must be considered in congruence. Part of the problem indicative with this is the notion of compartmentalized decision-making. It gives the impression that single or double combinations are acceptable. In sustainable decision-making, all three components are part of all decision-making. In much of the world, a fourth component – culture – is added as part of the Quad Stack (see

1.8 What Is Sustainability?

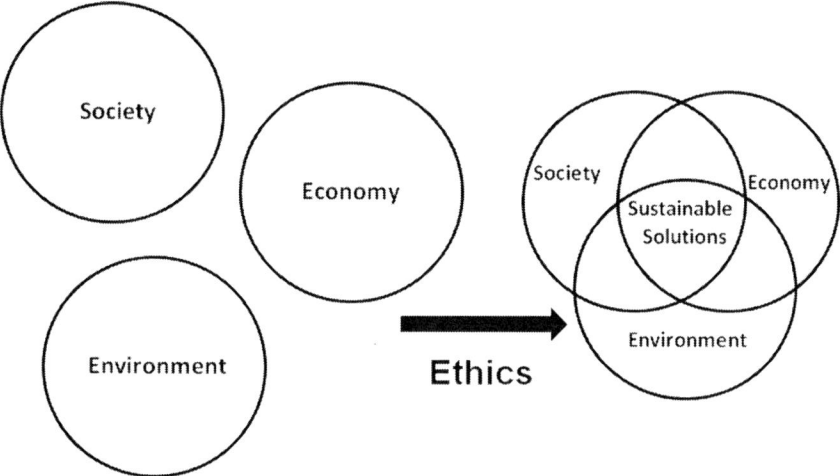

Fig. 1.5 The triple bottom line. Original graphic, Richard Jurin

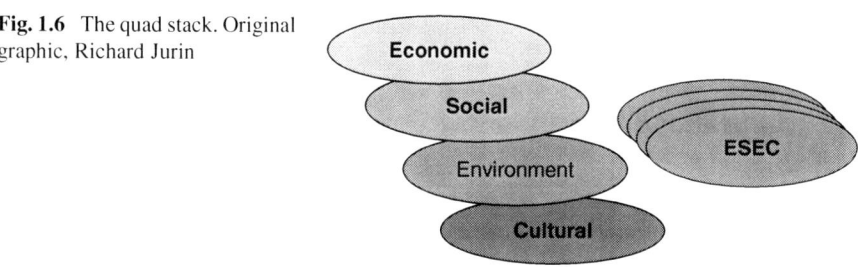

Fig. 1.6 The quad stack. Original graphic, Richard Jurin

Fig. 1.6). The actual 'stack' shows how all four components are considered equally in sustainability decision-making. A full, good quality of life includes more than just a high *standard of living* (*SOL*), a measure currently based solely on the health of the economy. GDP measures how well the economy is growing, which in its essence is about how much money is moving through a country's financial structure. GDP does not take into account that there are negatives and positives to how the money is moving. Nor does it factor how long-term factors (*externalities*) are contributing to decreased *quality of life* (*QOL*). Externalities are financial debits such as pollution produced during manufacturing a product that affects the health of people several years later through chronic exposure. GDP does not measure QOL, which is how people consider their own and society's well-being. When we do use measures, such as the Genuine Progress Indicator (GPI), the negatives and positives of financial growth are used. In general, the negatives are now outweighing the positives, such that even as the SOL still increases (as measured by GDP), the QOL is decreasing (as measured by GPI). Studies of well-being show that despite a threefold increase in SOL since 1957, we do not feel as though our lives are any better (Blackburn 2007).

As shown by the Triple Bottom Line and Quad Stack, a healthy, prosperous society relies on a healthy environment to provide food and resources, safe drinking water, and clean air for its citizens. These provisions are intertwined with QOL and the economic 'good growth' within the society. Simply put, American culture is usually equated to society, but they are two separate components in the rest of the world. Society can be equated to QOL while culture is a shared set of attitudes, values, goals, and practices within that group. Of course it is not that simple, but we are concerned here with communicating broad ideas, and we need to recognize that there are nuances we need to recognize when talking about people and communities. If we use a sustainability paradigm to understand well-being and QOL as our goal, then we see that environmental and socio-cultural factors cannot accept the inevitable negative consequences of current economic development paradigms. Sustainability offers a paradigm for thinking about a preferred future in which environmental, socio-cultural, and economic considerations are balanced in the pursuit of development and improved QOL.

1.9 Education for Sustainable Development

Many people see the term 'sustainable development' as a oxymoron because of negative, anti-environmental connotations of the word 'development.' The term 'growth' has similar responses since people think of urban growth and sprawl when they hear this word. Part of our job as environmental communicators is to clarify terms and show how they can be beneficial for future human living.

Instead of thinking of development and growth as sprawl within the context of sustainability, we need to reframe these terms. Development comes to mean mindful progress which enhances harmony between humans and the natural world, and growth gets redefined as becoming better, not bigger. As such sustainable development can now be seen as a term that means living better with well-being and an improved quality of life while still meeting the definition of sustainability. For communicators, the big question related to our work for sustainable development is, should we be attempting to influence behavior toward a predetermined and acceptable sustainable lifestyle? In other words, is our work part of civic engagement related to how we would like our societies to develop? If we answer affirmatively, much of what we do as communicators is about persuasion. Perhaps, the best way to persuade someone is by letting them make their own mind up through the use of fair and accurate information from a critical perspective.

1.9.1 Case Studies: Education for Sustainable Development (ESD)

Many universities worldwide are starting to consider the value of being academic centers of inspiration and expertise for the shift to sustainability. The Center for Environmental and Sustainability Education at the Florida Gulf Coast University works toward realizing this aspiration through scholarship, education, and action by involving the University, local community, and the wider community of scholars. The Center's goal of environmental and sustainability education advances understanding and achievement through innovative educational research methods, emergent eco-pedagogies, and educational philosophy and practice based on ethics of care and sustainability. The work of students is vital and includes youth activism, scholarly and editorial activity, planning Center events, and participation in green building and solar farm projects. It also provides opportunities for faculty, administrators, and staff to engage in scholarship, teaching, and service related to environmental and sustainability education. They promote a vision of leadership, scholarship, and action that is low budget but with high visibility. The University is constructing a 2 megawatt solar array occupying 16 acres. Although the project will reduce campus carbon emissions, some consider the solar array controversial in that it takes up ecologically sensitive land. The Center runs a monthly newsletter and has developed a student associates program. Such projects secure high levels of commitment from students, faculty, and community partners in promoting ESD.

Sustainable Development is a central thrust of many governmental, inter-governmental and non-governmental organizations, such as the United Nations, OECD (Organization for Economic Cooperation and Development), and the World Commission on Environment and Development. The United Nations has declared 2005–2014 as the Decade of Education for Sustainable Growth (SDE). According to OECD (Report: Supporting quality teaching in higher education, June 23, 2009), educational institutions can play a major role in contributing to sustainable development.

The Universiti Sains Malaysia (USM) has initiated a series of programs with support from the government to promote sustainable development within the country. These programs are (1) Kampus Sejahtera (or Healthy Campus) programme – a contextual framework of people within sustainability focused on the Quality of Life issues. Realigning of a new mission based on people led harmony and peace using concepts of indigenous wisdom; (2) *The University in a Garden* – a flowering of the mind metaphor. Working with USM students to improve community well-being and improve lives through an understanding of QOL. This work was begun with a student initiated ban on styrofoam use on campus; and (3) A transdisciplinary approach in promoting teaching and research activities to promote community action, and eventually to serve as a Regional Centre of Expertise on Education for Sustainable Development. This was a designation by the United Nations University in 2005, with five key thrusts: to move up the value chain, raise the capacity for knowledge, address persistent economic issues, increase QOL and sustainability, and strengthen the country and its infrastructure by becoming more sustainable with continuing improvements.

Credit: CESE logo. Center for Environmental and Sustainability Education, College of Arts & Sciences, Florida Gulf Coast University.

References and Further Reading

Altman I, Setha M (eds) (1992) Place attachment. Plenum, New York
Baskin Y (1997) The work of nature: how the diversity of life sustains us. Island Press, Washington
Blackburn WR (2007) The sustainability handbook: the complete management guide to achieving social, economic and environmental responsibility. Earthscan, London
Brundtland GH (1987) Our common future, report of the world commission on environment and development. United Nations, New York Oxford
Cantrill J (1999) The environmental communication commission. Ecologue Summer:1
Cohen MJ (2007) Reconnecting with nature: finding wellness through restoring your bond with the earth. Ecopress, Lakeville, MN
Corbett JB (2006) Communicating nature, how we create and understand environmental messages. Island Press, Washington
Cox R (2006) Environmental communication and the public sphere. Sage, Thousand Oaks, CA
Cox R (2007) Nature's 'crisis disciplines': does environmental communication have an ethical duty? Environ Commun 1(1):5–20
Daily GC (1997) Nature's services: societal dependence on natural ecosystems. Island Press, Washington
Day BA (2001) The new field of environmental communication. Appl Environ Educ Commun 1:1
Day BA (2002) Communication, a weapon in the fight against environmental health hazards. Appl Environ Educ Commun 1(1):1–2
Depoe S (2007) Environmental communication as nexus. Environ Commun 1(1):1–4
Diener E, Biswas-Diener R (2008) Happiness: unlocking the mysteries of psychological wealth. Blackwell, Malden, MA
Fallows D (2008) Almost half of all internet users now use search engines on a typical day. Pew Internet & American Life Project, http://www.pewinternet.org/cited 22 Apr 2009
Fortner RW ed (1991) Special earth systems education issue introduction. Sci Activities 28:1
Foulger D (2004) An ecological model of the communication process. Paper presented at the international communication spring meeting, New York
Kasser T, Kanner A (2003) Psychology and consumer culture: the struggle for a good life in a materialistic world. American Psychological Association, Washington
Kellert SR (1993) The biophilia hypothesis. Island Press, Washington
Lesk M (1997) How much information is there in the world? Paper presented at the Getty information institute time and bits: managing digital community meeting. Los Angeles, CA, 1998
Lyman P, Varian HR (2000) How much information? 2000, http://www.sims.berkeley.edu/how-much-info, cited 22 June 2009

Lyman P, Varian HR (2003) How much information? 2003, http://www.sims.berkley.edu/how-much-info-2003, cited 22 June 2009

McKibben B (2006) Age of missing information. Random House, New York

Meisner M (2008) What is environmental communication? (version 2.0). Environmental Communication Network, http://www.esf.edu/ecn, cited 30 December 2008

Mott FL (1939) A history of American magazines, vol 1. Harvard University Press, Cambridge, pp 1741–1850

Nature (1869) A weekly illustrated journal of science. Nature 1(2):ii

Neuman WR et al (2009) Tracking the flow of information into the home: an empirical assessment of the digital revolution in the US from 1960–2005. Paper presented at the 59th annual conference of the international communication association, Chicago, 2009

Ninan KN, Steiner AA (2009) Conserving and valuing ecosystem services and biodiversity: economic, institutional and social challenges. Earthscan, London

Pleasant A et al (2001) The literature of environmental communication. In: Aepli MF et al. (eds) Proceedings of the 6th biennial conference on communication and environment, Cincinnati, 2001

Pool IS (1983) Tracking the flow of information. Science m221(4611):609–613

Roush D (1998) Magazines as a medium for environmental communications. In: Senecah S (ed) Proceedings of the fourth biennial conference on communication and environment, Syracuse, 1997

Schoenfeld C (1969) What's new about environmental education? J Environ Educ 1(1):1–4

Schoenfeld C (1979) An annotated bibliography of environmental communication research and commentary: 1969–1979. Education Resources Information Center, Columbus, http://www.eric.ed.gov/cited 30 Mar 2009

Schoenfeld C (1981) The environmental communication ecosystem: a situation report. Education Resources Information Center, Columbus, http://www.eric.ed.gov/cited 30 Mar 2009

Schoenfeld AC (1982) American magazines and the environmental movement: symbiotic relationship, 1966–1975. Education Resources Information Center, Columbus, http://www.eric.ed.gov/cited 22 June 2009

Schoenfeld AC (1983) The environmental movement as reflected in the American magazine. Journalism Quart 60:470–475

Senecah S et al. (2004) Introduction. In: Senecah S et al. (eds) The environmental communication yearbook vol 1. Lawrence Erlbaum, Mahwah, NJ

Slosson EE (1922) Science from the side-lines. Century Ill Mag 107:471–476

Steele F (1981) The sense of place. CBI, Boston

Thayer RL (2003) Life place: bioregional thought and practice. University of California, Berkley

Wikipedia (2008) Environmental communication, http://en.wikipedia.org/wiki/Environmental_communication, cited 2 Jan. 2009

Wilson EO (1984) Biophilia. Harvard University Press, Cambridge

Witt W (1973) Communicating information model. J Environ Educ 5(1):58–62

Witten IH, Gori M, Numerico T (2007) Web dragons: inside the myths of search engine technology. Morgan Kaufmann, Amsterdam

Chapter 2
Communicating About the Environment

2.1 Introduction

When a natural resource professional says 'environmental communication,' what do they mean? The concept couples two terms, both of which are probably familiar. Still, both of them encompass large areas of meaning. We list a few definitions of 'environmental communication' and made a case for our own distilled version. Pulling apart the full term and attempting to state concise yet conclusive definitions for 'environmental' and 'communication' runs a high risk of failing. But, not defining one's terms carries an even higher risk of causing confusion later.

First, by 'environmental,' we defer to a nearly poetic definition put forth by Schoenfeld (1969), with input from a panel of ecologists. Here are their rhythmic, colorful phrases:

> In locus, the fouled, clogged arteries of the city quite as much as scarred countryside.
> In scope, a comprehensive, interrelated humankind-environment-technology system.
> In focus, global environmental impacts of crisis proportions threatening the well-being of all humankind on an over-crowded planet.
> In content, tough ecological choices, not easy unilateral fixes.
> In strategy, long-range impact analyses and rational planning.
> In tactics, grass-roots participation in resource policy formation – in the streets and through institutional channels.
> In prospect, a necessary reliance on alternative sources of energy.
> In philosophy, a commitment to less destructive technologies and less consumptive lifestyles.
> In essence, a recognition of pervasive interdependencies, that everything is connected to everything else.

As for 'communication,' the most popular American dictionary, *Merriam-Webster*, gives this meaning: 'a process by which information is exchanged between individuals through a common system of symbols, signs, or behavior.' Communication seems the more intuitive concept of these two terms. But definitions only take you so far in grasping a concept.

2.2 Communication Modeling and Theory

Understanding the process of communication can be enhanced through the use of models – graphic representations of some phenomenon. Numerous models explain how messages are sent and received and explain the many problems that can occur. The simplest model shows a sender selecting a channel through which a message is then transmitted to a receiver. For example, say you need to get a message to a fellow dormitory resident and decide to walk down the hallway and talk to them. You decided here to use face-to-face communication, with your voice's sound waves as your channel. Alternatively, you might have chosen to use a telephone or email as your channel. In any case, the message you sent would be almost the same. Information would be transferred from you to your colleague.

This simple model addresses only one-way communication, however. It fails to take into account the dynamics of two-way communication, where senders and receivers switch positions within the model as they take part in dialogue. Such dialogue might be an interpersonal conversation, a telephone call, exchange of letters, trading email messages, or any other of a number of possible interchanges. The inherent alternation of roles in two-way communication produces another important feature within the system – feedback.

Feedback involves a communicator's review and evaluation of message receipt and decoding. Feedback permits messages to be improved, whether by selection of a new channel or modifications within the message. Even if a communicative interaction is not in real time (as it is with traditional mass media), a careful sender strives to ascertain if the message was received and clearly understood (Fig. 2.1).

The cyclic model presented here depicts a dynamic system with sender and receivers interacting, each in turn encoding, transmitting and decoding. Thus, the sender becomes the receiver and the receiver becomes the sender. As can be seen in this model, a need exists for the sender to send a message, and to encode that message, in a form likely to be understood by the receiver. A mode/channel is selected and a message sent. This message is mentally processed by a receiver, who then responds in some manner. A receiver then alternates and becomes sender. The system, thus, perpetuates itself. Or, more precisely, communicating is an on-going process.

Any communication process is imperfect, as noise permeates the entire system. This pervasive impurity works against comprehension of messages. Noise may be

2.2 Communication Modeling and Theory

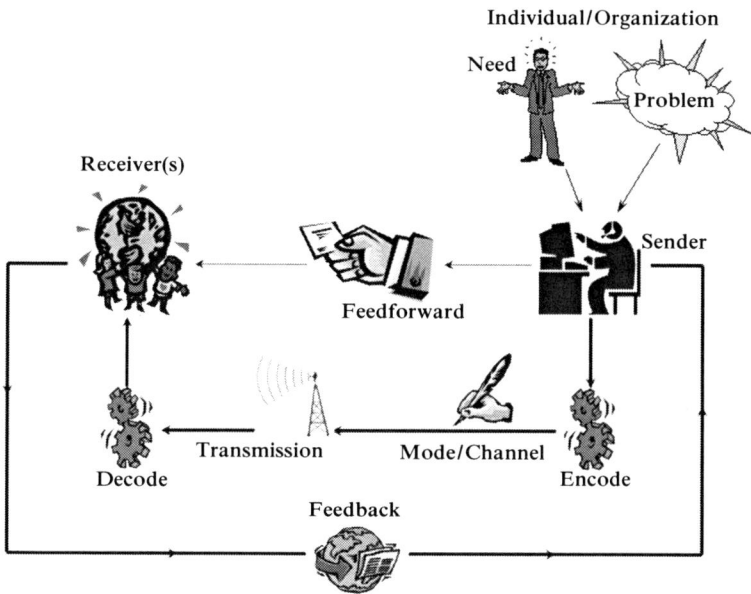

Fig. 2.1 A communication model. Original graphic, Richard Jurin

sounds unrelated to the message, but it can also be non-sounds. For instance, using engineering jargon in writing a children's book, providing a CD-ROM to a person without a computer, and showing a painting done only in subtle shades of pastel red, blue, and green to a color-blind person all would introduce lethal amounts of noise into a communication. The message would not get through. Anything present within a communication system that works against understanding is noise. Communicators strive to reduce it.

Closely related to feedback and helpful in overcoming noise, 'feedforward' allows communicators – both senders and receivers – to anticipate message exchange. You've probably experienced preparing for a complex session of communication. Studying for a test is one example. Deciding to see a movie based on a preview is another. When a sender forwards a simple prompt to ascertain if a receiver is willing and able to accept a message, this is feedforward, too. Feedforward allows a communicator to plan the expenditure of time and effort used in developing and interpreting messages. Mark Twain is credited with having once said, 'Tell 'em what you're going to tell 'em; tell 'em; tell 'em what you told 'em.' The first part of the adage is feedforward.

In examining models, there is a tendency to view the components as fixed and static, rather than as parts of a dynamic process. It is important to realize communication involves continually switching and changing of the sender-receiver roles. The model presented here is meant to guide you through the basic steps of the send-receive-repeat routine. Planning attractive and effective environmental communication relies on awareness of these modalities.

2.2.1 Communication Perspectives

Littlejohn and Foss (2008) survey communication theory through five basic perspectives. If we relate their perspectives to our basic communications model, we can see the complexity of communications and how it defies placement into a single category. All five perspectives have validity and give different views on how a message is created and modified by senders and receivers. Noise in the model can be equated to a breakdown in any of the five perspectives below.

The Mechanistic Perspective – this can be viewed as simply any message transmission from sender to receiver, and where feedback confirms a message was received intact without crippling noise.

The Psychological Perspective – here emotions and cognitive interactions of the receiver are prominent since these are related to correct decoding.

The Symbolic Interaction/Constructivist Perspective – this highlights the encoding, decoding and feedback parts of the model and assumes that in a successful communication the receiver and sender are co-creating the meaning inherent in the message.

The Systemic Perspective – this is a holistic, sum-of-the-parts consideration, because as the message is consistently moved from sender to receiver and then through feedback, unusual interpretations can act as noises to change the message in subtle ways – especially as more people become a part of the system.

The Critical Perspective – communications can be a source of empowerment and disempowerment for social groups. Is the message meant to coerce with a goal to either subjugate the receiver's thinking or to motivate the receiver with an aim to critical thinking?

2.2.2 Message Elaboration

Petty and Cacioppo (1981, 1986) developed the idea of 'message elaboration' where any message is cognitively processed by either a peripheral or a central pathway. For maximal impact, a message needs to be processed through the central pathway in which the receiver is motivated to scrutinize the message closely. After a great deal of thought, the message is committed to stronger neural retention pathways within the brain. A message that resonates strongly in the receiver is more apt to make a strong impact and ultimately produce a major positive attitude change in relation to the message. If, however, the message is seen as not having enough merit – the argument is too weak – it will be readily rejected after short consideration. The sender therefore must give the receiver a reason and a motivation to centrally process the message. It should induce some aspect of critical thinking.

On the other hand, an irrelevant or common-place message will be briefly (if at all) thought about and either be readily rejected, or have only a short-term and fragile effect with the sender's desired outcome. Message quality, credibility, attractiveness, importance, unique relevance (if the receiver thinks they already know about the subject of the message), or some aspect of noise may relegate the message to a peripheral pathway.

Receivers must be given ample motivation and have the ability (actually be capable of critical evaluation) within the message to determine the amount of elaboration they will give to the message. Of course, there is potential that any message might have components that induce either of the pathways separately in different individuals.

2.3 What Are the Differences and Similarities Among Environmental Communication, Environmental Education and Environmental Interpretation?

To this point, we've considered 'communication' as a basic function of being human. And, we've developed a concept of 'environmental communication' as evolving from an instinctual desire we have to understand the world around us. We create meaning from sensations and then converse about these meanings with others. Within this domain of endeavor, three related and overlapping fields have been established: environmental communication, environmental education, and environmental interpretation.

Try not to be confused by the redundancy of having environmental communication as the overarching concept, as well as one of the three fields within. 'Communication' covers a lot of conceptual territory. We contend that you must communicate to educate and to interpret. But, even though it is circular, you must communicate to communicate, too. Let us explain more.

The fields of environmental communication, education, and interpretation (ECEI) can be likened to a mythical monster: Cerberus, the canine guardian of Hades. This devil dog purportedly has three heads atop three necks protruding from a common body. The heads, though able – at least in legends – to attack different targets, were essentially the same creature. A natural resource professional can anticipate needing to call on communicative, interpretive, and educational skills daily in their on-the-job tasks.

So, the terms environmental communicator, interpreter, and educator can, for most intents and purposes, be used interchangeably. Yet, there are distinctions between the three. The following discussion delineates between the three fields, while also emphasizing their common features. A handy way to distinguish between environmental communication, education, and interpretation uses institutional setting of the informational exchange and audience focus to sort between formal, informal, and nonformal (Mocker and Spear 1982).

Formal education, by and large, takes place inside classrooms. Classroom-based institutions, from pre-schools through universities, are formal educational centers. In formal settings, students are usually evaluated on criteria based in specific lesson objectives. Learning is achieved through interaction with a teacher. Although some relaxed teaching methods may permit substantial interactive dialogue among teachers and students, learning is highly structured in formal settings. Most often, learners submit to the educational goals of the institution and are required to attend formal education classes. They are a captive audience. The educational institution holds the decision-making power over what is learned and where learning takes place. *Environmental education occurs in formal settings.*

Informal education features less structure, no formal evaluation (no required tests of knowledge), and no requirement of attendance by learners. They attend because they are interested and/or desire entertainment. Institutions of informal education include museums, parks, nature centers, wildlife refuges, zoos, aquaria, art galleries, and historical sites. Learners can elect to attend informal educational programs when visiting an informal educational facility. Learning is often facilitated by a professional, perhaps a naturalist, docent, or living history actor. Interaction is loose compared to formal education, and learners are encouraged to ask questions freely. There exists for the visitor a personal option to interact with the interpreter. Informal learning needs to be entertaining because it is a leisure-time activity for visitors. The educational institution decides what is learned about, but cannot dictate whether learners attend to the lessons being offered. *Environmental interpretation occurs in informal settings.*

Nonformal education involves information disseminated primarily through mass media – television, radio, newspapers, pamphlets, fact sheets, billboards, magazines, the Internet, etc. These communications can transcend place and time. Even though you've may have never actually been to the Amazon rainforest, the Siberian steppe, or Antarctica, you probably still have some idea what each looks like and the environmental issues that affect these places. Messages in the mass media provided you with that information. Learning via mass media is controlled by the learner, who decides unilaterally what to pay attention to. Nonformal learning is also a leisure-time activity, though it may carry undertones of requirements for job or school. By picking and choosing what mass media messages to subscribe to and commit to memory, the learner controls both what is learned, when to learn it, and where learning takes place. They are only limited by the amount of information given out by the medium to which they pay notice. *Environmental communication occurs in nonformal settings.*

Each of these three settings – formal, informal, and nonformal – depends on a mediator. This mediator may be called a teacher, facilitator, host, interpreter, journalist, communicator, tour guide, information/outreach specialist, educator, scientist, engineer, or one of many other titles. The environmental communicator's job is to transmit environmental, scientific, and/or natural resource information to interested recipients.

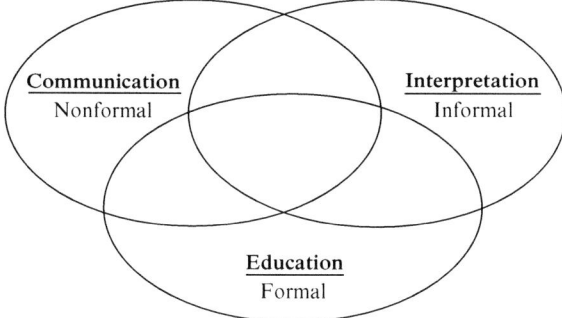

Fig. 2.2 Overlapping relationships of environmental communication, education, and interpretation, for the professional. Original graphic, Richard Jurin

Self-directed learning is a fourth type of possible setting, where the learner discovers information through first-hand experience, and occurs outside of ECEI. This is because self-directed learning is not within the control of any mediator or institution. No professional helps the learner in any way in self-directed learning.

In Fig. 2.2, overlapping circles signify that communicators/educators/interpreters will find themselves involved in doing two or even all three types of activity as the situation commands. Their audiences have subtly different motivations which will help a professional pick techniques from environmental education, interpretation, and communication. Techniques and resulting ideas can be ascribed to each, any, or all of the components. In many cases the distinction between the components may be difficult to see.

2.4 Principles of Adult/Community Education

Adult education is the discipline of *Andragogy* – the teaching and education of adults. Adult-only audiences are generally more focused in what they want to learn and how they wish to learn it, than children. Adult education is frequently non-formal and informal, and includes learning new skills and learning for personal development. In this sense, adults are rarely a captive audience and thus response best when they are motivated to see value in the learning. Adult experiences and accumulated knowledge can be a boon in getting them motivated, but also problematic in that they may have acquired much misinformation that needs to be overcome.

Community education, also known as Community-based education and Community learning, is similar to adult education but tends to focus more on community groups trying to resolve community needs and problems through joint action. Sometimes a community may not openly admit or even realize they have an issue (as in many environmental-related problems); they do not want to be dictated to by outsiders. As such, communities may need some low-key guidance in helping to define their own terms and solutions, rather than being told what they are lacking. Community members usually know what they want in improved quality of life, but may lack the knowledge, skills, and resources to make informed decisions. In general, community education works best when it helps community members to be self-determinant about identifying their needs and problems; are motivated and encouraged about their own capacity to resolve the situation; use community services and resources, which are inclusive of all members of the community; and have responsive non-authoritative officials from governmental and agency community support systems.

2.5 Implications for the Professional

Communicators, interpreters, and educators need to understand each and every context in which they hope to transmit their environmental information. A single topic will have to be developed and packaged in different ways for different audiences. To summarize, here's a final splitting of these fields:

Environmental education

Mode of education	Formal – tightly structured learning with specific objectives
Main media	Personal presentation, structured interaction
Main focus	To teach
Primary interaction	One-on-group
Audience	Not volunteers
Main institution	Schools

Environmental interpretation

Mode of education	Informal – loosely structured learning/augmented incidental learning
Main media	Personal narration, interactive dialogue, interpretive signage/displays (exhibits, trails, etc.)
Main focus	To entertain
Primary interaction	One-on-one, or one-on-small group, or directed self-guidance
Audience	Volunteers
Main institutions	Historic centers, museums, wildlife refuges, art galleries, zoos, museums, parks, nature preserves, etc.

2.5 Implications for the Professional

Environmental communication

Mode of education	Nonformal – incidental structured learning
Main media	Mass media (newspapers, magazines, brochures, pamphlets, fact sheets, displays, TV, radio, Internet) – transcends place and time
Main focus	To inform
Primary interaction	Through audience size and reaction to advertising – purchasing power
Audience	Volunteers
Main institutions	Mass media companies

Regardless of all the parsing we could do to differentiate between environmental communication, education, and interpretation, they remain so closely related that if they were organisms they would certainly be classified as the same species. Consider some of the largest professional organizations of environmental communication professionals:

National Association for Interpreters (NAI)
International Association of Public Participation (IAP2)
Society of Environmental Journalists (SEJ)
North American Association for Environmental Education (NAAEE)

NAI has an environmental education section. IAP2 has members in academia, industry, government, as well as free-lance writers and artists. NAAEE has a section devoted to different groups of nonformal professionals. SEJ allows educators to become members. It's fair to say environmental professionals dealing with communication are not hung up on labels for themselves.

2.5.1 Case Study: Environmental Education. Project WILD

In 1983, Project WILD debuted as a softbound book of lesson plans about wildlife, targeted to elementary school teachers. In order to receive a curriculum guide, teachers participated in a workshop, lasting from a couple of hours to a couple of days. Developed by wildlife agency personnel in 12 Western U.S. states, Project WILD has grown steadily, currently reaching all 50 U.S. states, the District of Columbia, Puerto Rico, plus Canada, Japan, India, Czech Republic, Iceland, and Sweden. Project WILD, as an organization, publishes six curriculum guides and additional supplementary materials, has state and national-level staff, and conducts on-going evaluation to assure educational value. Project WILD integrates science, language arts, social studies, math, art and other disciplines, and is used by resource specialists, naturalists, rangers, and youth leaders, in addition to K-12 teachers. The delivery model pioneered by Project WILD – upfront, short-term training required to receive materials – is emulated by many EE programs. Project WILD assisted in the development of Project Learning Tree (about trees and forests) and Project WET (about water). Together, these three programs have been called 'EE's Big Three.' In 2006, Project WILD trained its 1,000,000th educator, making it the most widely-used environmental education program in the United States and, possibly, the world.

Credit: Project WILD K-12 curriculum and activity guide cover (Council for Environmental Education)

2.5.2 Case Study: Environmental Interpretation. Great Barrier Reef Marine Park

Located in the Pacific waters off the east coast of Australian state Queensland, the Great Barrier Reef Marine Park is defined by superlatives and impressive environmental interpretation. The park is 2,500 km long, covers 345,400 km^2, and contains about 2,900 coral reefs plus 940 islands. It is a National Park, a United Nations World Heritage Site, and the largest coral reef system in the world. Great Barrier Reef is visited by 4.9 million people annually, including almost 2 million tourists, about half of those international visitors (Hariott 2002) A 25-year strategic plan, completed in 1994, guides all human activities. To manage recreation and tourism in the park, the Australian government – via a standalone agency, the Great Barrier Reef Marine Park Authority – collects an 'Environmental Management Charge' from visitors using any of the 820 permitted tourism operators. The fee has generated about AU$7 million per year since 1993 and supports an array of educational and interpretive services, including traditional media such as signage, rangers, fact sheets, and newsletters, plus more elaborate means including cutting-edge web sites and ReefHQ, an aquarium containing 4 million liters of coral creatures on display. A national sense of responsibility toward the reef is strong. Nearly all operators have adopted voluntary, strict best practices, which includes providing educational talks on the way to the reef. A standard fact sheet was meticulously written, designed, and tested before being distributed to as many park visitors as possible (Moscardo 1999).

Credit: Children at Reef HQ. *Credit*: Commonwealth of Australia.

2.5.3 Case Study: Environmental Communication
An Inconvenient Truth

A 2006 documentary film about global warming, *An Inconvenient Truth* features former U.S. Vice President Al Gore. The film posits global warming as genuine, potentially cataclysmic, and primarily human-caused. After leaving elected office in 2001, Gore returned focus to climate change, finding and revising a PowerPoint presentation he had created while advocating for a national carbon tax and the 1997 Kyoto Protocol to the United Nations Framework Convention on Climate Change. Dubbed 'the slide show,' Gore gave his presentation 1,000 times before producers Laurie David and Lawrence Bender and director Davis Guggenheim worked to transform the presentation into a film. Interestingly, around the same time in 2004, Gore hired Duarte Design, of Mountain View, CA., to juice up his presentation by condensing content and using multiple new technologies, moving 'the slide show' into an Apple-computer-based Keynote package allowing for inclusion of high-definition video and complex animations. On film, the pumped-up 'slide show' is woven with stories from Gore's life, to powerful effect. The film succeeds by several measures. Commercially, *An Inconvenient Truth* cost $1 million to make yet earned $49 million at the box office. It earned two Academy Awards ('the Oscars') and played a large role in Al Gore's receipt of the 2007 Nobel Peace Prize. In impact, the film generated ample civic discussion about climate change's severity. For example, Belgian activist and housewife Margaretha Guidone convinced 200 high-ranking government officials, including her country's prime minister, to watch *An Inconvenient Truth*.

Credit: Al Gore speaking about '*An Inconvenient Truth in Tokyo*'. *Credit*: Yugi Ohugi/WireImage/Getty Images.

References and Further Reading

Beck L, Cable TT (2002) Interpretation for the 21st century: fifteen guiding principles for interpreting nature and culture, 2nd edn. Sagamore Publishing, Champaign, IL

Boyte HC (2005) Everyday politics: reconnecting citizens and public life. University of Pennsylvania Press, Philadelphia

Boyte HC, Shelby D (2008) The citizen solution: how you can make a difference. Historical Society Press, Minnesota

Corbett JA (2006) Communicating nature: how we create and understand environmental messages. Island Press, Washington

Cranton P (2006) Understanding and promoting transformative learning: a guide for educators of adults, 2nd edn. Jossey-Bass, San Francisco

Dainton M, Zelley ED (2004) Applying communication theory for professional life: a practical introduction. Sage, Thousand Oaks, CA

Griffin E (2008) A first look at communication theory. McGraw-Hill, New York

Fazio JR, Gilbert DL (2000) Public relations and communications for natural resource managers. Kendall/Hunt, Dubuque, IO

Ham SH (1992) Environmental interpretation. North American Press, Golden, CO

Hariott VJ (2002) Marine tourism impacts and their management on the great barrier reef. Cooperative Research Center for the Great Barrier Reef Technical Report No. 46, Townsville, QLD, Australia

Hartman LA, Hanna JD (1987) Interpretive educational courses in the United States and Canada. J Environ Educ 18(4):1–7

Heimlich JE (1993) Non-formal environmental education: towards a working definition. ERIC Clearinghouse, Columbus, OH

Jacobson SK, McDuff MD, Monroe MC (2006) Conservation education and outreach techniques. Oxford University Press, Oxford

Jacobson SK (2009) Communication skills for conservation professionals, 2nd edn. Island Press, Washington, DC

Kleis RJ (1974) Non-formal education: the definitional problem. Program of studies in non-formal education discussion papers number 2

Knowles MS, Holton EF, Swanson RA (2005) The adult learner: the definitive classic in adult education and human resource development, 6th edn. Butterworth-Heinemann, Massachusetts

Littlejohn SW, Foss KA (2008) Theories of human communication, 9th edn. Thomson Wadsworth, Belmont, CA

Longo NV (2007) Why Community matters: connecting education with civic life. State University of New York Press, Albany, NY

Machlis GE, Field DR (eds) (1984) On interpretation: sociology for interpreters of natural and cultural history. Oregon State University Press, Corvallis, OR

Merriam SB, Caffarella RS, Baumgartner LM (2006) Learning in adulthood: a comprehensive guide. Jossey-Bass, San Francisco

Merriam SB, Brockett RG (2007) The profession and practice of adult education: an introduction. Jossey-Bass, San Francisco

Mocker DW, Spear GE (1982) Lifelong learning: formal, non-formal, informal, and self-directed. ERIC Clearinghouse, Columbus, OH

Monroe MC (ed) (1999) What works: a guide to environmental education and communication projects for practitioners and donors. New Society Publishers, Gabriola Island, Canada

Moscardo G (1999) Communicating with two million tourists: a formative evaluation of an interpretive brochure. J Interpretation Res 4(1):21–37

Mullins GW (1984) The changing role of the interpreter. J Environ Educ 15(4):1–3

Parker L (2008) Environmental communication: messages, media & methods, 2nd edn. Kendall/Hunt, Dubuque

Petty RE, Cacioppo JT (1981) Attitudes and persuasion: classic and contemporary approaches. William C. Brown, Dubuque, IA

Petty RE, Cacioppo JT (1986) Communication and persuasion: central and peripheral routes to attitude change. Springer, New York

Pierssene A (1999) Explaining our world: a guide to environmental interpretation. Taylor & Francis, London

Radcliffe DJ, Colletta NJ (1989) Nonformal education. In: Titmus CJ (ed) Lifelong education for adults: an international handbook. Pergamon Press, Oxford, pp 60–63

Schoenfeld CA (1969) What's new about environmental education? J Environ Educ 1(1):1–4

Tilden F, Craig R, Bruce C, Dickenson RE (2008) Interpreting our heritage, 4th edn. University of North Carolina Press, Chapel Hill

Umphrey M (2007) The power of community-centered education: teaching as a craft of place. Rowman & Littlefield, Lanham, MD

Vella J (2002) Learning to listen, learning to teach: the Power of Dialogue in Educating Adults. Jossey-Bass, San Francisco

Wlodkowski RJ (2008) Enhancing adult motivation to learn: a comprehensive guide for teaching all adults. Jossey-Bass, San Francisco

Chapter 3
Developing Your Environmental Literacy

3.1 Introduction

Environmental literacy denotes an individual's set of abilities and commitments necessary to find, understand, assess, and act on information about the health of our environment. So, environmental literacy embodies values, beliefs and attitudes toward sustaining a healthy environment. Prerequisite to environmental literacy is a standard conception of literacy and a more specific idea, science literacy. An environmentally literate person understands the workings of modern science, and of policy-making. They also know how to apply their abilities to affect changes in society. Each of these elements builds on the others. Being aware of and having knowledge about environmental problems only supports the higher order skills necessary to be fully environmentally literate.

To understand environmental literacy, we must first understand literacy and, then, science literacy. Literacy encompasses some form of competency, whether in literature, cooking, yachting, child rearing, playing the violin, or any other recognized vocation. At the core of literacy is the ability to read and write so as to learn from the knowledge of others and then contribute to any particular body of human endeavor. In addition, there is also a sort of literacy concerned with numbers and their application, called numeracy. As the sheer amount of data created and quick access afforded it, mostly through Internet, has expanded massively, so has need for knowledge and skill in mathematics, logic, and statistics.

A person communicating environmental information needs to be able to judge accurate and relevant scientific information and data, and relate it in credible ways to broader, mostly non-scientific audiences. Likewise, to be able to judge what human activities are sustainable requires the communicator to know about the functioning of the environment, to be sensitive to extra ecological pressures placed on the environment by modern humans, and to know what constitutes wise decisions. The communicator then attempts to help others go through the same thought processes. In doing so, environmental communicators assist others in becoming environmentally literate. This chapter discusses literacy, science literacy, environmental literacy, and, finally, the arts of filtering knowledge and argumentation.

Knowing how all these notions interact informs the communication process. You'll then be better able to develop clear and credible messages for all the audiences you'll need to reach as an environmental communicator.

3.2 Literacy

Being able to read and write is indispensable for citizens seeking to function well in modern societies. The new service-based economies of post-industrial, developed nations place greater value on higher levels of literacy. This implies a continuum along which an illiterate person gains skills in reading and writing, and subsequently may attain a threshold level that allows them to be literate enough to contribute to society. In a Western society, this means people are comfortable with certain functions such as being able to read and fill out a tax document, read and understand a newspaper, follow instructions from product labeling, navigate the Internet to find desired, truthful information, and comprehend dosage a drug prescription. Young children and adults who never learned to read and write would be at the lower end of the continuum. Toward the upper end would be people who are the most gifted readers, writers and critical thinkers.

Miller (1989), in a groundbreaking study, reported one-quarter of the United States populace was either illiterate or functionally illiterate. If valid, this finding presents a huge problem for any communicator when developing any message for non-specialist audiences in the United States. The US census defines illiteracy as anyone above the age of 14 that possesses less than a fifth-grade education. People with literacy deficiencies have difficulty coping with common day-to-day chores. Consider:

- Twenty percent cannot read a bus schedule or address a letter.
- Understanding a typical insurance policy requires a 12th grade reading level.
- Reading the instructions on over-the-counter medicines takes a 10th grade reading level.
- Making a TV dinner or filling out a tax form requires an 8th grade reading level.
- Comprehending data on a driver's license uses a 6th grade education.

Remediation of reading problems is the focus of many in-school and adult education programs. About 10% of those 65 million Americans who lack literacy are enrolled in some form of remedial education program. That's a start, but it is not a final solution to allowing all to participate fully in the American dream of democratic participation in society.

3.3 Numeracy

The United Kingdom has a national strategy for developing numeracy (Department for Education and Employment 1999), stated as:

> Numeracy is a proficiency which is developed mainly in mathematics but also other subjects. It is more than an ability to do basic arithmetic. It involves developing confidence and competence with numbers and measures. It requires understanding of the number system, a repertoire of mathematical techniques, and an inclination and ability to solve quantitative or spatial problems in a range of contexts. Numeracy also demands understanding of the ways in which data are gathered by counting and measuring, and presented in graphs, diagrams, charts and tables.

This explanation has been reproduced in educational policy around the world. Recently, for example, Malta adopted it verbatim within that country's 'National Policy and Strategy for the Attainment of Core Competences in Primary Education' (Ministry of Education, Culture, Youth and Sport 2009).

Numeracy, like literacy, has many factors. Comfort with numbers large and small, a variety of operations involving them, logic, and mathematical reasoning are but a few. Beyond basic math, distinct areas of geometry, algebra, statistics, probability, and modeling present information about the environment and its condition. Natural resource professionals rely on numeracy no less than other number-hungry job-holders such as architects, financiers, and business owners. Within the natural resource professions, engineers and theoretical biologists possess high numeracy. Managers who deal daily with both budgets and data from the field are no less reliant on this type of proficiency. Communicators working side-by-side with managers, scientists, and engineers may not achieve the same heights of numeracy, but they need to know enough to ask appropriate questions about numbers, so as to elicit salient explanations and be able to produce clear communiqués. Intriguingly, university-trained journalists have been found sorely lacking in numeracy, with one study finding 58% of news directors said newly-degreed job applicants were unable to understand a municipal budget (Tomorrow's broadcast journalists 1996). In a more recent study, skill with numbers did not enter the picture painted of self-reported qualities present in award-winning broadcast journalism students (Edwards et al. 2005).

Perhaps, it should Frankel (1995) wrote, 'Deploying numbers skillfully is as important to communication as deploying verbs.' We agree.

3.4 Science Literacy

Science is a complex subject, which requires a special set of analytical skills to understand. In that scientific findings play a large role in policy-making, scientific literacy is important for democratic functioning. To attain science literacy, citizens need to (Miller 1989):

- Understand the vocabulary of science.
- Understand the scientific method, the process that differentiates science from other ways of knowing.
- Comprehend research's iterative practice that views knowledge as tentative for centuries.

- Possess critical reading skills which enable one to judge valid and reliable scientific findings from incorrect, biased, or misleading ones. Part of this requirement is an understanding of the peer-review process, wherein research reports are reviewed by other scientists before being published or presented publicly.
- Analyze the costs and benefits of science and technology on the global village.
- Realize each new piece of knowledge adding to the overall picture of the universe's function. Therefore, few studies are ever really 'breakthroughs.'
- Further realize uncertainty is always present in scientific knowledge. So, it should always be reported.

Even experts rarely have a foolproof answer to any problem. Understanding of natural phenomena is imperfect. Science seeks to build consensus, not prove this or that. Science is like an ever-expanding jigsaw puzzle. As pieces are found and the picture nears completion the borders expand. New information is constantly required (Bauer 1994).

The relationship between standard literacy and science literacy can be envisioned as the intersection of two continuums. In Fig. 3.1, the horizontal line represents literacy, one's reading and writing competency. The vertical line signifies science literacy.

Look at the four quadrants formed by the lines. Individuals and groups of people can fall into any of the four quadrants. An excellent communicator would be able to craft messages which could be understood by persons in all of these categories. Be aware the type of message should by necessity be different for each quadrant's audience, and the medium may differ to reach each quadrant.

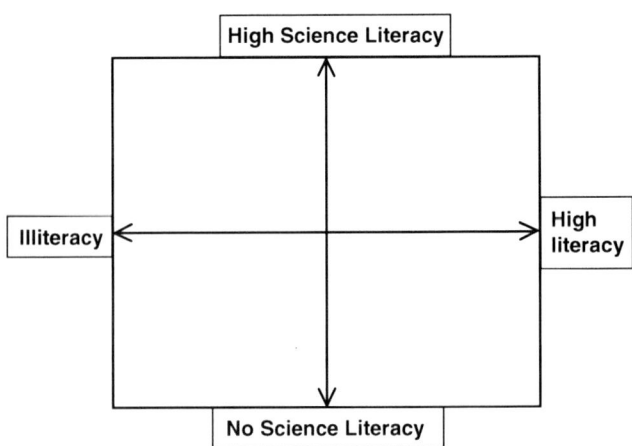

Fig. 3.1 Relating literacy as different from science literacy. Original graphic, Richard Jurin

3.5 Environmental Literacy

The modern built environment with its cities, highways, agricultural fields, and factories has increased human potential to affect planet-wide ecology. We now have the power to degrade our entire biosphere. Environmental literacy acknowledges this power and saddles it with responsibility to focus on restoration, conservation, and sustainability. This responsibility enfolds all human discourses about our interrelationships with the environment. An environmentally literate person is able to perceive and interpret the relative health of environmental systems and to take appropriate action to maintain, restore, and improve the health of those systems. A prerequisite to environmental literacy is being scientifically literate. Orr (1992) notes environmental literacy carries within it a quality of mind that seeks connections and draws from intimate experience with at least one ecosystem. So, working knowledge of ecology is critical for anyone hoping to make good decisions about sustaining the environment.

The concept of environmental literacy implies (Roth 1992):

- Knowledge of environmental issues and the credible science behind them.
- Understanding of the 'whole picture' and not just minor parts of it.

    Empathy toward the total environment.
    Knowledge of action skills.
    Environmentally responsible beliefs, values and attitudes.
    Willingness to invest personally.
    Active involvement in solving environmental issues.

3.5.1 Degrees of Environmental Literacy

As with literacy, numeracy and science literacy, environmental literacy varies by degrees. That is, environmental literacy also lies along a continuum. Three stages have been outlined (Roth 1992). They emphasize benchmarks along segments of the continuum.

Nominal environmental literacy (at the lower end of being environmentally literate)

- Developing awareness and environmental sensitivity
- Increased respect for nature and concern for how humans interact with it
- Rudimentary knowledge of natural systems and how humans interact with them

Functional environmental literacy (in the middle reaches of environmental literacy)

- Broader knowledge and increased understanding of human/environment interactions
- Awareness and concern about negative human interactions with environment

- Developed skills with which to analyze, reason and evaluate environmental information
- Ability to communicate conclusions and feelings about problems/issues to others
- Willingness to act to resolve problems/issues of personal concern

Operational environmental literacy (at the higher end of environmental literacy)

- Broad and deep knowledge of ecology and human–environment interactions
- Routine evaluation of environmental impacts of human actions and consideration of their consequences
- Active and regular gathering and evaluating of relevant information
- Making decisions among alternatives, followed by advocating appropriate actions
- Holding a strong sense of responsibility and personal investment for the environment
- Acting at several societal levels, from local to global, as well as both personally and collectively
- Living an ecologically sustainable lifestyle

Note these characteristics of environmental literacy apply to technologically developed societies, such as the United States and European Union. Many of the world's people do not live in such industrialized countries. Many peoples who live in pre-industrial or underdeveloped societies are in a qualitatively different situation. While it may be argued indigenous peoples 'live in harmony' with the environment, this appears most true when they live without modern technology and with minimal contact with industrialized values. Our purpose here is not to assign blame for environmental degradation, however. Suffice it to say environmental degradation exists, and good communication practices can help to foster operational environmental literacy, through which sustainability by definition results.

Roth (1992) concludes environmental literacy is the goal of all environmental education. Given you necessarily must communicate to educate, here's a direct corollary: *Environmental literacy is the goal of all environmental communication.*

3.5.2 Measuring Environmental Literacy

Education reform, a restructuring of formal institutions especially those delivering Kindergarten through high school instruction, rests heavily on two pillars: standards and assessments of learning. Educational standards are stated expectations of learning, presented prior to the start of an episode of teaching. Assessments, often in the form of standardized tests, are meant to measure learning against standards and provide accountability for the educational institution. Now about 25 years down the reform road, transformation to standards and standardized assessments has been rocky.

3.5 Environmental Literacy

Given that environmental literacy is an even-newer construct, to no surprise, only a few educational standards address it head-on and attempts to measure it are pioneering yet imperfect. Among the United States, Pennsylvania has a separate set of enacted 'environment and ecology' academic standards, Wisconsin has 'model academic standards for environmental education' though these fall outside core disciplinary areas, and California has mandated environmental education content in their science frameworks. When it comes to measuring environmental literacy, the U.S.'s National Environmental Education & Training Foundation created a 9-year series of annual public polls, with the help of the Roper Reports (Coyle 2005). Annually, 1,500 randomly-selected American adults were surveyed on their environmental knowledge and behavior. Overall results were decidedly mixed:

- Only 12% of Americans passed a basic quiz on energy topics.
- Just one third of Americans knew what a 'watershed' was.
- Two thirds believe environmental protection and economic development are compatible.
- Ninety-five percent of Americans support education about the environment in public schools.

Coyle (2005) concluded only 1% or 2% of Americans could be declared environmentally literate, based on knowledge and skills measured by the polls.

Globally, the Program for International Student Assessment (PISA) carries out worldwide assessments of academic achievement for 15-year-olds every third year. PISA's parent organization, Organisation for Economic Co-operation and Development (OECD) has 30 member countries and an additional 28 nations were included in the 2006 testing when PISA focused on science competencies, including a gamut of environmental and natural resource questions. PISA 2006 may offer, as OECD claims, 'the first comprehensive and internationally comparative database of students' knowledge about the environment and environment-related issues' (Green at fifteen? 2006), making it a valiant attempt at measuring environmental literacy in teens around the world.

What does 'Green at Fifteen?' show?

Teenagers taking part were mostly well aware of environmental issues, but often knew little about their causes, raising questions about how well we will be equipped to tackle challenges of the future. More than 90% were familiar with issues relating to air pollution, nuclear waste and water shortages. But, almost half were unable to identify a single source of acid rain. The top average score came from Finland. Taiwan had the highest proportion of students at high proficiency (Fig. 3.2).

Overall, widespread environmental awareness among teenagers was coupled with a sense of responsibility and optimism. In contrast, the results showed a lack of realistic appreciation on the part of lower-scoring students for effort and time needed to address environmental problems. If, as tomorrow's voters and taxpayers, they remain unconvinced of the scale of the challenges, they will be unlikely to be ready to bear the cost of forward-looking investments in this arena. There is a risk that ignorance may lead to complacency and inaction.

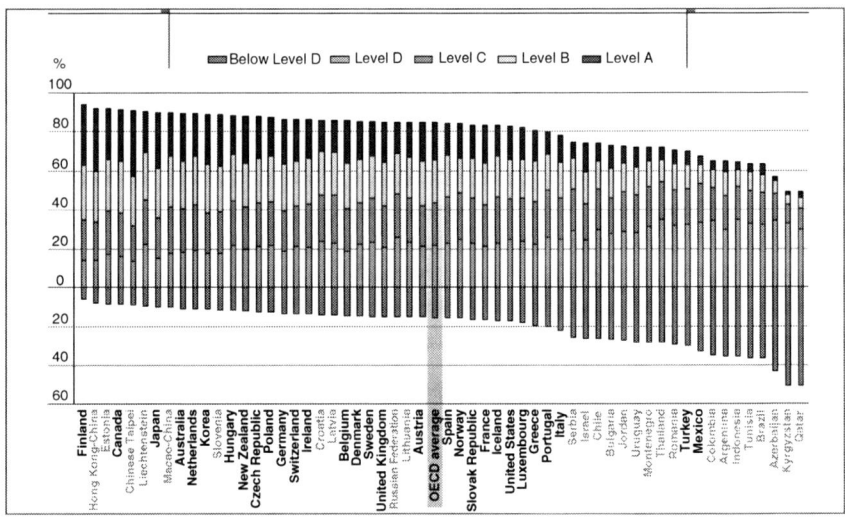

Fig. 3.2 International environmental science performance index. *Source*: Green at Fifteen?: How 15-Year-Olds Perform in Environmental Science and Geoscience in PISA 2009, p. 38, www.oecd.org/pisa

On average, about one fifth of each country's students were able to tackle the hardest environment-related science questions, such as finding alternative explanations for an increase of CO_2 emissions and the rising temperature of the earth. At the other end of the scale, however, an average 16% of each country's students were unable to cope with basic environment-related questions, such as how freezing water can contribute to soil erosion.

Higher-performing students said they used mass media and the Internet to find out about environmental issues. For the bulk of students, school was the most common source of information. On a sobering note, students with the best grasp of environmental science were the least optimistic that things will improve in the future.

3.6 Ecological Literacy

David Orr (1992) and Fritjof (1997) argue there's a major component of literacy missing. They see a need to add a heavy dose of systems thinking to Environmental Literacy. Systems thinking is the ability to understand the interactive complexity of natural systems. They call the resulting concept 'Ecological Literacy,' as it concerns the principles of organization of ecosystems and their application in understanding how to build a sustainable human society. We work with nature at all levels in this integrated and holistic way of thinking. We view the complex *interdependency* of all life forms and their reliance on *abiotic systems*. How systems are organized and interact becomes more crucial than a reductionist view of components in isolation.

As a connection to literacy, Goleman (2009) identifies *ecological intelligence* to explain why we consumers are victims of a 'blackout of information' concerning the full life cycle cost and economic pricing of the products we buy. We are oblivious to the full system costs of manufacturing processes from mining through processing, manufacturing and all the waste streams within the process. He does emphasize, however, that the new information age is about to bring us 'radical transparency' and so herald in a new era where the informed consumer is sovereign in their buying decisions, which will help promote true 'green' business practices.

3.6.1 Nature-Deficit Disorder

When children are kept indoors more than 95% of their lives, they are hard pressed to develop empathy and knowledge of natural environments even those right outside their own doors and windows. They become detached from ecosystems and nature's services. They fail to see connections between abiotic and biotic components, they are pre-empted from thinking systemically, and have only ephemeral experiences in natural settings. Louv (2005) coined the term 'nature-deficit disorder' to capture this disconnect between children and nature. What's more, he notes at least one nature-deficient generation has grown into adults and carry the burdens of the deficit with them. These burdens are 'human costs of alienation from nature' not limited to 'diminished use of the senses, attention difficulties, and higher rates of physical and emotional illnesses' (Louv 2005, p. 34). Social trends to blame are parental fears arising from ever-more gruesome news stories and overestimation of risks, a lack of access to natural areas rooted in liability concerns and litigation, and steeply increased screen-time in front of televisions and computers. Rampant childhood obesity and rising rates of childhood depression and Attention Deficit-Hyperactivity Disorder proffer corroborating evidence of an enormous and troubling social problem.

Nature-deficit disorder severely hampers, if not fully blocks, development of environmental literacy. Environmental communicators, educators and interpreters talk about their work as an 'antidote' or 'cure' for nature-deficit disorder.

3.6.2 No Child Left Inside

Nature-deficit disorder put resonant words onto cultural concerns noticed by many, though not a proportionally large segment of the American population. The idea galvanized educators, environmental organizations, worried parents, and policy-makers, adding needed stickiness to a message environmental communicators had been trying to get out for at least a couple of decades. Charles et al. (2008) state 'the message that getting children outside is a prerequisite for happy, healthy children is universal, and the broad emotional resonance of the issue is a powerful tool we have at our disposal.'

A resultant national policy from this movement is U.S. legislation called No Child Left Inside. This legislation, to be an amendment to the huge Elementary and Secondary Education Act (the law governing federal government involvement in K-12 formal education), would return environmental education to classrooms where it has been pushed out by education reform efforts. No Child Left Inside has three components:

- Funding to train teachers in environmental education and to operate model programs, especially those with outdoor learning.
- Funding to each state which submits a complete environmental literacy plan, to ensure high school graduates are environmentally literate.
- Grants at national, state, and local levels to build capacity to facilitate environmental literacy.

Like no legislation before, No Child Left Inside would integrate environmental education across the American curriculum.

3.7 How Science Information Becomes Reliable

Being literate hinges on knowing the credible, valid, reliable, and trustworthy from that which is not. The Internet being likened to an 'information superhighway' suggests incredible accessibility via a fast and vast corridor. But, a communicator needs to be quick to discern good information from not-so-good information, since both can appear on your computer screen with the same ease. Knowing how information becomes credible and acceptable within the scientific community can help.

Core to acceptance of research findings by scientists is the process of peer review. 'The Scientific Knowledge Filter' (Fig. 3.3) show the winnowing of information by peers who judge its merits based on the quantity and quality of other reviews it has undergone in the community of science. The knowledge filter strains out non-scientific knowledge. Information which cannot be verified and does not fit a consensus of the science community, after being reviewed by more experts, is discredited or removed. This is not meant to imply that all such knowledge is without merit. Indeed, emotions, spirituality, folklore, and intuition are vital components of human culture. They are not science, however.

At the top of the filter information generated by human endeavor is input. This input includes misinformation, misconceptions, folklore, and pseudo-sciences such as astrology, psychic phenomena, and fad dieting. Material from such sources does not fit the consensus of reality held by modern scientific thinking. In most instances, they cannot be demonstrated by repeatable observation and measurement. This removes them from the realm of actual science. Still, these pursuits are not completely discounted as there are ways to convincingly validate them to many. There is enough circumstantial evidence from 'believers' to continue their existence in the popular culture. When was the last time you read your horoscope?

3.7 How Science Information Becomes Reliable

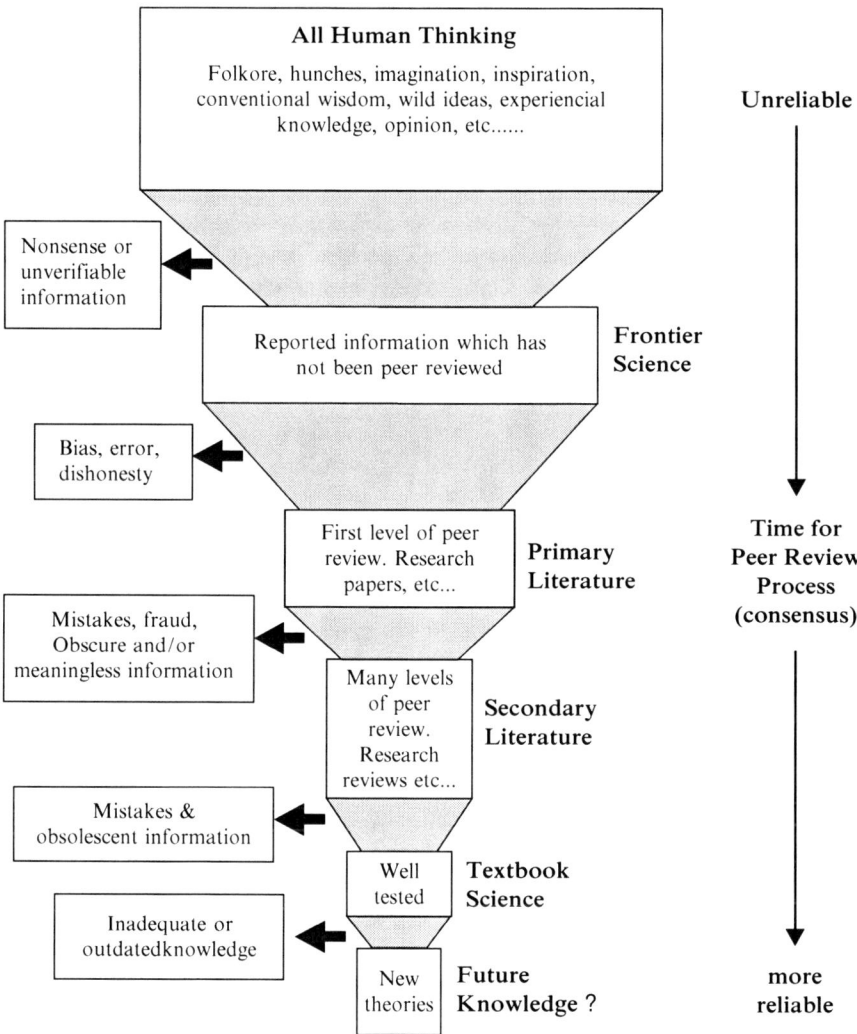

Fig. 3.3 The Scientific knowledge filter. Henry Bauer, Scientific Literacy and the Myth of the Scientific Method, 1992

Each subsequent step of the knowledge filter further screens ideas, hypotheses, contentions, and speculations. Credibility rises as the amount of unverifiable knowledge decreases. At each level more experts, who have extensive experience with similar information, review the remaining material and judge it to fit existing ideas or declare it erroneous. We have applied labels to various steps along the way.

3.7.1 Frontier Science

After draining off nonsense and information that can never be verified, the knowledge filter continues to hold information that meets the most cursory criterion of seeming to be scientific. At this stage, this 'scientific' material appears without the benefit of peer review. An example might be scientists who announce major findings through press conferences, in order to generate wide public interest in their work. Sometimes this is done as a tool to increase funding for further research, often focused on outlandish concepts. Many documents of frontier science are written and published by self-proclaimed experts. These books are generally not peer-reviewed. To the undiscerning reader, this information represents expertise. In reality, it should be judged with heightened skepticism.

3.7.2 Primary Literature

In the first stage of scientific peer review, a report is reviewed primarily for its face validity, guarantees of reliability, and contribution to a particular field. Face validity means that it holds together and seems to make sense during a close reading. Reliability deals with stating the methods used and showing how the data were acquired and handled. To be reliable a study has to offer instructions for others to repeat the work. Because science is iterative, and builds on itself, a study most always connects with others. This connection may be to confirm or contradict other findings.

Primary literature includes research journals, scientific reports, academic books, and conference proceedings. Findings released at this filter level may, after closer inspection, be declared to have flaws in methodology, analysis, and interpretation. Fraud is also a possibility against which science is constantly vigilant. Consensus builds when data and results can be replicated and begin to make more sense when fitted into current theories. Material that is flawed or does not fit current theory is usually rejected, at least until more corroboration can be offered. It is worth noting that many revolutionary ideas in science have been initially rejected, only to become future benchmarks. Revolutionary science, however, tends to be hotly debated before it is tentatively accepted by a minority in the science community, after which it then makes its way down the filter slowly.

3.7.3 Secondary Literature

Scientific material that has been replicated and fits the orthodox views of the science community finds its way into the secondary literature. This material is composed of larger reviews and monographs written by experts in specific disciplines, as well as media focused on bridging science and the wider public.

Examples include *Popular Science* magazine, the Discovery Channel on cable and satellite television, and books by authors such as Jared Diamond (e.g., *Guns, Germs, and Steel*), Barry Lopez (*Arctic Dreams*), and Malcolm Gladwell (*The Tipping Point*). They strive to synthesize many studies, fitting all the data into a larger picture. Much of this material may have first been published several years prior to being reviewed. Scientists and science journalists who undertake reviews spend large amounts of time contemplating a body of work, evaluating with unusual rigor though not usually generating their own data on the matter at hand.

3.7.4 Textbook Science

Scientific theory-building begins to take shape between secondary literature and this level of the knowledge filter. After continuous acceptance during several years of replication and further study, valid and reliable ideas make their way into textbooks, compilations of the best works in particular fields. The next generation of investigators learns from these books. In such books, theories that form the knowledge structures of science are explained and purported to describe reality with parsimony.

As information passes through the knowledge filter, it gradually becomes more accurate as findings exhibiting bias and lack of rigor are filtered out. Time information spends within the knowledge filter increases as information moves down each step of the way. Note also that consensus is based on the prevailing theories and models within each scientific discipline. Still, even highly developed concepts at the textbook level of the filter can be invalidated by newer thinking that develops better theories. For instance, plate tectonics overwrites continental drift in geology during the middle of the twentieth century. Such throwing out of once strongly held theory occurs most often in rapidly developing fields, such as astronomy, quantum physics, and environmental science. In each of these cases, new instrumentation and techniques have recently uncovered previously unknown data.

An unfortunate attribute of the knowledge filter, information most readily available to non-scientific public is generally from nearer the top of the filter. This material is the most likely scientific information to catch the attention of the gatekeepers of the mass media, and these news outlets are where most adult Americans learn about new scientific discoveries (Miller 1989). Contentiousness between journalists and scientists is something environmental communicators are apt to have to deal with on a regular basis (Friedman et al. 1986). A time and place where the contents of scientific textbooks make the news is almost assuredly far, far away.

3.7.5 The Internet and the Knowledge Filter

The internet has created a situation that effectively bypasses this kind of knowledge validation by peers and 'experts.' Anybody can, and many do, publish their ideas and work directly to the internet usually through personalized web-pages.

Originally it was thought that open access to everyone would provide better peer review. Except for a few specialized sources, such as certain government science/engineering sites and science sites that do have validation processes in place, this is unfortunately not true for the bulk of information on the internet. Many of the peer review journals we talk about in the filter now have electronic versions, but nevertheless still have peer review as part of the publication process. Unfortunately for the bulk of information on the internet, the user must be savvy enough to understand if the information is valid or simply conjecture based on somebody's opinion, which may be part of the top of the filter process we describe above.

If you 'Google' a topic, you will most likely find one of your first results listed is Wikipedia. Many students often use this as the only source of information, which can be a dangerous thing. On entirely non-controversial material the entry may be factually correct in the main. Wikipedia does have a vetting process where other readers can edit an entry. It is thought that after a while with enough people editing the entries, this will create reliability within the site – this assumes that the 'editors' actually know enough to be experts in the material they are editing. Sadly, a lot of material, especially in the environmental arena, is not neutral and is directly affected by strong and usually conflicting opinions making valid and reliable information hard to discern. Even entries in online dictionaries can have glaring errors. In Chapter 9 we cover internet sources more and offer guidelines of how to gauge reliability about a websites information. Using the internet is a good start, quick and efficient, but think about the bottom of the knowledge filter towards peer reviewed information – the library is still your best source of reliable information, at least for the final check.

3.8 Thinking Critically About Scientific Information

Critical thinking helps distinguish between facts and opinion. It means subjecting all facts and conclusions to careful and reasoned analysis. Scientists collect data and then use it to generate information that then forms into new knowledge. The means through which they generate data, information and knowledge is the scientific method. Knowing how research is conducted allows a critical reader of scientific reports to ask the right questions in order to judge the worth of the study in their own search for environmental solutions. The following set of questions can guide a critical analysis of scientific information.

3.8.1 Does Their Argument Make Sense?

What is the crux of their argument? What are they claiming to have found? Many times a clear sounding statement such as '80% of the population of the United States are environmentalists because they have environmental attitudes' really tells us nothing useful. What is an environmental attitude? Is it one attitude or many specific attitudes that defines being environmental? Who was doing the defining? How was the 80% figure derived? We often have to make inferences and decisions based on scarce evidence. Avoid 'leaping to conclusions' by critically reviewing the information. Get to the core of their argument by closely scrutinizing the abstract and the discussion sections of the report.

3.8.2 Who Is the Source of the Information?

Examine who is making claims, and what their motivations can be inferred to be. Try to expose hidden agendas. This does not mean that either an environmental group's or an industry's positions are automatically biased beyond acceptance. Everyone can be considered a member of some special interest group or another. Call on your own reserve of knowledge, but also look for the source to offer alternative points of view or contrary evidence. Knowing what makes a source of information reliable helps to weed out misinformation and propaganda.

3.8.3 Are the 'Facts' Placed in a Context of Accepted Knowledge?

Facts are always contextual, in that they must be understood within a bigger picture. Blanket statements need to be supported with information. Think of a statement claiming, 'The number of cancer patients in America today is double what it was per capita when compared to 50 years ago.' Before we conclude our modern pollution problems are responsible, we should first ask what background information is being left unmentioned? What average ages are cancer patients today compared to before? Is there an increase of older patients with cancer? What are the main causes of death today compared to before? Are third world nations developing heavy industry now showing the same trend? Modern medicine is improving so that we now have a population that is not dying from relatively minor diseases. Has this new trend been figured into the statement? Such questioning leads to a richer understanding of any subject.

3.8.4 How Was the Information Obtained?

We are bombarded with information all the time. It often sounds like 'Chicken Little' is always predicting the sky is about to fall. When it doesn't, we are not surprised anymore. We are becoming desensitized to the myriad information we now receive. Questions such as 'Says whom?' 'Who paid?' and 'What then?' get at the ways and means by which the information was gathered and then sent to you.

This line of inquiry is especially critical in instances of 'one study panic.' A startling example of this happened in 1989, when many people in North America were scared with a single report that Alar, a pesticide used on apples, was found in apple juice given to children. At the time most of the studies about Alar had shown it to be relatively harmless to humans, yet this report claimed Alar was detrimental to children. A major environmental group and a major TV network championed a crusade against Alar manufacturers and apple growers. The one contrary study was eventually dismissed by the U.S. Environmental Protection Agency, but not before a lot of damage to the apple growing industry. The moral: view findings in context with what is already known.

3.8.5 What Kind of Study Was Reported?

Most research studies can be grouped as either correlational or experimental. Did the study look at something that already existed and make conjectures about the outcome, or did it control the variables and show the outcome as a cause-effect relationship?

Correlational research makes educated guesses that certain observations are linked in some logical way. For example, if a small town's residents begin to have an increased number of cancer reports and at the same time the groundwater in the area is found to contain traces of the carcinogen benzene, then it might be conjectured the two are linked. Since there is no tight control of all the variables that might have produced an outcome, further studies are usually necessary.

Experimental research controls all factors to isolate the one or more identified variables (the cause) that force an outcome. Most drug and chemical studies are done in this way so that results seen are attributed to the experimental compound alone. A test population, often cell cultures or non-human animals, is given different doses of a chemical and then observed to see when the chemical becomes toxic, and how the toxicity is manifested in observable symptoms. Later clinical studies in human volunteers are used to gauge reliability of drugs.

3.8.6 Were Measurements and Statistics Used Properly?

Statistics are complex mathematical calculations that indicate the significance of data. Significance here means statistically sound, showing a difference highly unlikely to be chance, and not necessarily important. But, misuse and erroneous reporting of statistics is cause for concern. If an amount of a chemical found in a lake is significantly higher than it was a year ago, should we be concerned? Researchers refer to this as the 'So what?' question. Finding data that reaches statistical significance is one thing, determining if it matters in the larger scheme is another.

Another important statistical consideration is the size of the study population. Though it is expensive to run test on many subjects, small populations are more difficult to generalize from and introduce more uncertainty into outcomes.

In summary, critical thinking means we need to look at the big picture and examine all data and methods closely. We need to look for hidden or obscured causes and effects, and unstated assumptions. We certainly need to avoid illogical thinking where simple answers are expected or are convenient. We need to question our own deeply help assumptions and biases. In short, we need to know as much about an issue as possible before rendering any opinion or decision and to remain as open-minded as possible, even when the outcome may be contrary to what we may have originally desired. To do this, communicators need to learn how to suspend judgment, consider the preponderance of evidences, and construct tenable arguments for our audiences.

3.9 The Art of Argumentation

Asking insightful questions allows you to dissect arguments contained in research reports. As an environmental communicator, be aware that astute members of your audience will be doing the same to your messages. Constructing solid arguments becomes indispensable to your success.

The art of argumentation is one of the most ancient skills attributed to learned persons. Rhetoric has a legacy that spans dozens of centuries. As we can see from the last section, asking the right questions is critical. Rhetoricians seek to build invincible arguments. Environmental communicators can borrow some rhetorical techniques to aid in the presentation of well-reasoned dispatches.

In stating and supporting a position – what we mean by arguments here, as opposed to anger-sparked tirades – your messages have to be able to withstand three essential questions that will be asked by critics:

- What do you mean?
- How do you know?
- What was presupposed?

Bear these three deceptively simple questions in mind as you build your messages. Close and careful consideration of these questions will focus your attention on your presumptions, evidence, inferences, and the manner in which you combine them.

- **Presumptions** A presumption is a statement of fact or belief for which no verification is required. These are technical points of agreement accepted by participants as true. Usually any one objecting to a presumption has the 'burden of proof' to show it untrue.
- **Evidence** One selects bits and pieces of reality to support one's claims. These items are evidence. Evidence substantiates and attempts to verify what is being claimed. Empirical assertions have factual origins. Value assertions are drawn

from aesthetic judgments. Policy assertions are based on ethical expectations. Each can be a source of evidence.
- **Inferences** Inferences are derived from your evidence and are based on your presumptions. They are conclusions about what is believed to be correct. Reasoning is used to reach specific determinations. This reasoning can be deductive, where a specific point is reached from examining generalities; inductive, where a generalization is made based on specifics; or a combination of both. All inferences should be relevant and logical.

Arguments use a number of language devices to structure their content. Argumentative structure deals with the manner presumptions, evidence, and inferences are put together and presented to the audience. The most common language devices are analogy, metaphor, simile, and example.

- **Analogy** To compare two things, which are alike in some respect, so as to explain a concept is to make an analogy. These extended explicit comparisons are common and useful to argument-makers. The familiar is used to explain the difficult to understand, highlighting similarities while downplaying or ignoring dissimilarities. An example might be that attempting to communicate to a hostile audience is like stepping into the lion's cage. It emphasizes that a lot of care and preparation needs to be in place before the communicator can expect the audience to be receptive.
- **Metaphors** Metaphors equate one presumably familiar thing with another presumably less familiar one. Because they have great power in initial explanations, metaphors are abundant in everyday language. Yet, most metaphors exhibit vagueness and ambiguity under close examination. When a singular metaphor is used extensively, it tends to break down and loose its explanatory effect. An example, 'the communicator is molding an audience's opinion as a sculptor chisels a block of granite.' The idea of crafting messages to develop a persuasive outcome seems fine, yet people are not inert pieces of rock. The rock is not influenced by anything but the sculptor. People are influenced all the time by factors outside the communicators control, hence the metaphor breaks down under scrutiny.
- **Similes** Similes compare a familiar thing with something less well-known, just as a metaphor. They, by definition, use of the word 'like' in making the comparison. This preposition serves to both highlight and qualify the comparison. An example is a communicator who argues that a small community trying to work with a mega-corporation over a local pollution problem is like David confronting Goliath. It emphasizes that the size difference is daunting, yet like David, the community can get the mega-corporation to listen to their grievances if the community squarely faces the problem and remains focused.
- **Examples** When one thing is used to represent a group, an example is being made. Examples are inductive devices because a sample is used to show the validity of many. In explaining complex issues, examples are extremely useful, almost necessary, to the staking a clear position. An example of this is when a previous oil tanker spill off a coastline is used to argue for establishment of regulations for all oil tankers in coastal waters.

Validity of any argumentation is only as good as the rigorousness of the presumptions, evidence, inferences, and language devices being used to construct it. Sound construction of an argument results from sound reasoning. Arguments are weakened by unsound components and faulty construction. Such weaknesses are referred to as logical fallacies. Argumentation experts (e.g., Makau 1990; Freeley, 1997) list fallacies as falling within language, evidence or reasoning of an argument.

Language

- *Ambiguity* – more than one interpretation can be applied to a premise.
- *Vagueness* – meaning is too inexact to contribute to the discourse.
- *Equivocation* – changing meaning during a discourse to make an argument seem more compelling than it is.
- *Obscuration* – hiding behind needless jargon, terminology, semantics, etc.

Evidence

- *Repeated assertion* – giving an argument over and over again in hopes it will become more acceptable.
- *Non-representative instance* – using a poor example.
- *Insufficient instances* – not giving enough examples.
- *Invalid statistical measure* – using biased, atypical samples, and other inappropriate mathematics and statistics to bolster an argument.
- *Unreliable source* – using biased, non-credible, or inappropriate sources.

Reasoning

- *Straw argument* – using weak versions of opposing or alternative views.
- *Begging the question* – avoiding or circumventing the relevant issue.
- *Circularity* – using unsupported assertions or simply restating a claim using different wording.
- *Non sequitur* – stating an irrelevant claim that does not follow the argument's evidence.
- *Appeal to ignorance* – using untenable burden of proof to force acceptance of an assertion.
- *Appeal to popular prejudice* – relying on what most people think to force acceptance of a claim.
- *Appeal to tradition* – relying on conventional social practices to enforce the correctness of a position.
- *Ad hominem* – challenging the maker of the claim instead of the claim itself.
- *Over-simplification* – overlooking potentially relevant information in order to make the issue easier to understand.
- *Hasty generalization* – using small, biased, untypical samples as evidence for a broader group.
- *Post hoc* – erroneously trying to emphasize an alleged cause-effect relationship.
- *Faulty comparison* – drawing conclusions from unwarranted comparisons.

Environmental communicators will continually be constructing arguments. While there are many pitfalls to avoid, there are also many salient positions in need of defending. Indeed, concepts within operational environmental literacy call on practitioners to take firm stances in restoring, conserving, and sustaining healthy ecosystems. Nevertheless, taking a position and making a stance need not be contentious. One can cooperate and still make a sound argument.

Cooperative argumentation focuses on reasoned interaction about a controversial issue, with the intention of helping participants make the best assessments and decisions possible under the given circumstances. Such a process usually leaves participants feeling better about the decisions made, since all involved have shared information and ideas. Such buy-in makes for longer lasting solutions to issues.

Counter to cooperation, competitive argumentation focuses primarily on winning a debate. It may be referred to as 'combative interaction.' This process alienates participants rather than clarifying a situation. Each participant concentrates on trying to either prove themselves right or to prove their opponents wrong.

Arguments can be competitive or cooperative, inclusive or combative. Participants in an argument most always have power to decide which sort of interaction they wish to have. Allies and enemies can both be made through argumentation. Environmental communicators are wise to take a long-range view and strive for cooperation over competition, to make allies instead of enemies.

3.10 Case Study: Environmental Literacy. Last Child in the Woods

Louv's *Last Child in the Woods* documents the marked decrease in time American kids spend in direct, unmediated contact with nature. As a science journalist, he noticed trends of children spending less time outdoors, of the erection of legal fences around areas where children used to play, of swelling fright toward unlikely abductions, and of more and more addictive behavior toward video games and on-line sources of children's leisure. Louv spent 10 years considering evidence and collecting stories. He then penned a passionate, well-reasoned, insightful, and strongly argued book. *Last Child in the Woods: Saving Our Children from Nature-Deficit Disorder* hit the global marketplace of ideas in 2005. It is a powerful example of secondary literature getting compelling science into more hands and minds. In 'nature-deficit disorder,' environmental communicators found a rallying cry to popularize their work and to seek policy improvements. The No Child Left Inside Coalition and Children & Nature Network were both formed. Together, they represent more than 1,340 organizational members representing at least 50 million individuals. No

Child Left Inside, proposed federal legislation in the United States, would provide unprecedented support – to the tune of $100 million – to plan, produce and more accurately measure environmental literacy (Charles et al 2008).

Credit: Book cover 'From Last Child in the Woods, by Richard Louv. © 2005 by Robert Louv. Reprinted by permission of Algonquin Books of Chapel Hill. All rights reserved.'

References and Further Reading

Arons AB (1990) Achieving wider scientific literacy. Daedalus 112(2):91–122

Bauer HH (1994) Scientific literacy and the myth of the scientific method. University of Illinois Press, Illinois

Browne N, Keeley SM (2006) Asking the right questions: a guide to critical thinking, 8th edn. Prentice Hall, Upper Saddle River, NJ

Charles C, Louv R, Bodner L, Guns B (2008) Children and nature 2008: a report on the movement to reconnect children to the natural world. Children & Nature Network, Santa Fe, NM

Clayton CW (2007) The re-discovery of common sense: a guide to: the lost art of critical thinking. iUniverse Inc, New York

Coyle K (2005) Environmental literacy in America. National Environmental Education & Training Foundation, Washington

Department for Education and Employment (1999) The national numeracy strategy: framework for teaching mathematics for reception to year 6. Cambridge University Press, Cambridge

Diamond J (1997) Guns, germs, and steel: the fates of human societies. W.W. Norton, New York

Edwards DL, Tuggle CA, Kozlowski D (2005) How do we select them and then what do we teach them? A survey of success factors for student broadcast journalism award winners. Feedback 46(2):12–22

Elder JL (2003) A field guide to environmental literacy: making strategic investments in environmental education. Environmental Education Coalition, Rock Spring, GA

Frankel M (1995) Word and image: innumeracy. *The New York Times*, p 24

Freeley AJ (1997) Argumentation and debate: critical thinking for reasoned decision making, 9th edn. Wadsworth, Belmont, CA

Friedman SM, Dunwoody S, Roger CL (1999) Communicating uncertainty: media coverage of new and controversial science. Lawrence Erlbaum, Mahwah, NJ

Friedman SM, Dunwoody S, Rodgers CL (eds) (1986) Scientists and journalists: reporting science as news. AAAS, Washington

Fritjof C (1997) The web of life: a new scientific understanding of living systems. Anchor, New York

Gladwell M (2000) The tipping point: how little things can make a big difference. Little Brown and Co, Boston

Green at fifteen? How 15-year-olds perform in environmental science and geosciences in PISA 2006 (2009) Programme for international student assessment, Organisation for Economic Co-operation and Development, Paris

Goleman D (2009) Ecological intelligence: how knowing the hidden impacts of what we buy can change everything. Broadway Books, New York

Hamm CM (1989) Philosophical issues in education: an introduction. Falmer, New York

Hill B, Leeman RW (1996) The art and practice of argumentation and debate. Mayfield, Mountain View, CA

Hirsch ED, Kett JF, Trefil J (2002) The new dictionary of cultural literacy: what every American needs to know, 3rd edn. Houghton Mifflin Harcourt, Boston

Huff D, Geis I (1954) How to lie with statistics. W.W. Norton, New York

Johnson L (2005) Teaching outside the box: how to grab your students by their brains. Jossey-Bass, San Francisco, CA

Kaner S, Lind L, Toldi C, Fisk S (2007) Facilitator's guide to participatory decision-making. Jossey-Bass, San Francisco, CA

Lakoff G, Johnson M (1980) Metaphors we live by. University of Chicago Press, Chicago

Lopez B (1986) Arctic dreams. Vintage Books, New York

Louv R (2005) Last child in the woods: saving our children from nature-deficit disorder. Algonquin, Chapel Hill, NC

Makau JM (1990) Reasoning and communication: thinking critically about arguments. Wadsworth, Belmont, CA

Miller JD (1983) Scientific literacy: a conceptual and empirical review. Daedalus 112(2):29–48

Miller JD (1987) Scientific literacy in the United States. In: Evered D, O'Connor M (eds) Communicating science to the public. Wiley, London

Miller JD (1989) Scientific literacy. Paper presented at the annual meeting of the AAAS, San Francisco, 17 January, 1989

Ministry of Education, Culture, Youth and Sport (2009) National policy and strategy for the attainment of core competences in primary education. St Venera, Malta

Orr DW (1992) Ecological literacy: education and the transition to a postmodern world. State University of New York Press, New York

Paulos JA (1989) Innumeracy: mathematical illiteracy and its consequences. Hill and Wang, New York

Rieke RD, Sillars MO, Peterson TR (2008) Argumentation and critical decision making, 7th edn. Allyn & Bacon, Boston, MA

Roth CE (1992) Environmental literacy: its roots, evolution, and direction in the 1990s. ERIC Clearinghouse, Columbus, OH

Rybacki KC, Rybacki DJ (2007) Advocacy and opposition: an introduction to argumentation, 6th edn. Allyn & Bacon, Boston

Pauley J (1996) Tomorrow's broadcast journalists: a report and recommendations from the Jane Pauley task force on mass communication education. Society of Professional Journalists, Greencastle, IN

Walton DN (1989) Informal logic: a handbook for critical argumentation. Cambridge University Press, Cambridge

Chapter 4
Investigating Environmental Issues

4.1 Introduction

Environmental communicators find they spend a lot of time and effort considering proposed solutions to environmental problems and investigating environmental issues. Issue analysis is central to the making of an optimally environmentally literate person. An ability to understand the roots of environmental problems and issues better equips one to effectively communicate about how to resolve them. First, we need to distinguish between problems and issues.

Problems are smaller units of controversy, whereas issues most often are larger societal disputes. Though this distinction can be fuzzy, problems tend to conglomerate into issues. A problem can be thought of as singular and more likely to involve an us-against-them position of participants. They can be extremely difficult to solve, because of this adversarial stance between the players. Often two sides are forced into a winner-take-all situation where the losers get nothing. Not surprisingly, dealing with problems are can be combative and foster lingering antagonistic attitudes. Environmental problems such as whether to log a forest, strip-mine an area, or dam a river hinge on singular decisions. Once a decision is made, damage is done. Fortunately for communicators, because of their either-or nature, problems lend themselves to rapid understanding. Still, stakes can be high and outcomes costly to the loser.

Issues, on the other hand, are much more difficult to comprehend. They have many facets and aspects to be managed if they are to be resolved. Because issues are constituted from collections of problems, they can be termed multidimensional. They also demonstrate morphing at varying speeds. Some issues may swell, whereas others may dissipate. Examples of issues include global climate change, air and water pollution, land use management, endangered species, and human population growth. There are many, many others. Notice how the labels applied to issues give them a luster of scientific detachment and do not really describe how people feel about them.

Issues do not usually hinge on singular answers to questions. Instead, they tend to involve the answering of chains of questions. Each answer is commonly greeted by the arising of new questions. As a series of answers are reached, they drive management of the issue. Whereas problems dissipate after one decision, issues require strings of decisions to mitigate. Resolution of issues takes a long time and sometimes is never fully reached.

Consider some issue-linked questions: How should wilderness be managed? Should human population be managed? What constitutes acceptable levels of risk for industrial chemicals? How do we resolve conflicting interests for land uses? How can tropical rainforests be preserved? What are the most prudent uses of pesticides and herbicides in farming? Answering such questions taps directly into beliefs and values upon which positions are established.

Issues hold meaningful importance to individuals, groups, and communities. They have consequences, either real or imagined. Although hugely relevant in confronting issues, importance is extremely difficult to measure and not much easier to characterize. Describing what an issue means to a party comes from understanding the party's beliefs and values which underlay their position on the issue. And, there is always more than one position to consider.

Because individuals, groups, and communities start dealing with any issue from different places, different perspectives are inevitable. Commonly, there are several points of view on any particular issue. Over time, however, multiple points of view may devolve into polarized sides. Each of these two sides supports an often-oversimplified opposing theme, which remain after alternate minor points of view have been subsumed. Have you ever heard an issue summed up as 'jobs vs. the environment'?

Attempting to resolve issues tends to cut to the core of a person's belief and value systems, and so must be treated with great deference. The very act of challenging a position on an issue can be tantamount to challenging the person directly. *When beliefs and values are threatened, the result is outrage and fear.*

Outrage and fear further compound issues by making it more difficult for those with differing points of view to communicate. This fear-based type of noise can be severely damaging to any communication system. Systems can utterly break down when infected with fear. Outrage can also degrade into physical violence. Fist fights, terrorist acts, and even wars have come about when values and beliefs about environmental issues have been offended. Clear and cogent environmental communication early in an issue's emergence provides some insurance against such unfortunate outcomes.

4.2 Components of Issue Analysis

- **Problem** A situation in which someone or something is at risk. The problem is agreed upon, but dispute lies in deciding on a solution. Likewise, the process for finding a solution is likely to be debated.
- **Issue** A multidimensional situation bringing together a related set of problems, where different values and beliefs about the problems are held by various

players/stakeholders. Usually there is little consensus on how to prioritize and address the various dimensions within the issue. Key questions to elucidate an issue might be: What is at risk? Do all stakeholders stand to lose or gain equally? If not, what is at jeopardy for each? What does each stand to gain? What does each stakeholder hold most dear within the issue? What can be negotiated? Answers will vary based on differing perspectives of stakeholders. It is essential to define the situation clearly for all positions before proceeding to solutions.

- **Players/Stakeholders** Individuals, groups or organizations holding a vested interest in an issue and its outcome. What different roles are there among individuals, groups or organizations involved in the issue? Are there individuals, groups or organizations who might be involved, but who are not stakeholders? Can such an entity serve as a mediator or just complicate the group dynamics when trying to form a solution? What hidden interests exist to detract from the defined issue?
- **Positions** The stance or postures various players adopt concerning an issue. Differing values and beliefs exist for the various players, so their positions will differ too. Do you understand why the various players hold the beliefs and values they do? Are the players defending their positions based on valid information? Is the decision being made by all players an informed one based on reasoned thinking? Or, is additional information and re-clarification needed? Are all alternatives given critical review? Or, has reaching a solution, any solution, become the goal? This last condition is called 'groupthink,' a pathology where reaching consensus becomes tantamount to true analysis of alternatives.
- **Solutions** Alternative strategies and ideas employed to find consensus and resolution for an issue (Ramsey et al. 1989). Is a compromise among stakeholders possible? Is consensus possible? Do all alternatives address the defined issue and not just vested interests? Who will be affected by the short- and long-term consequences of a solution? What are the costs and benefits to all players associated with a solution? Will the solution work permanently?

4.3 How Issues Arise

Issues do not spontaneously arise in modern societies. They require champions who seek to amplify a set of concerns and highlight a group of problems. Issues arise because a group, or rarely a particularly vocal individual, decides to 'push' their agenda into wider recognition. Champions hope to gain attention, to build support and to drive expansion of concern. All these functions are based on communicating well.

Protecting the canyon lands and red-rock country of the Colorado Plateau as federally designated wilderness is the mission of the Southern Utah Wilderness Alliance (SUWA). The group formed in 1983 and seized on a deceivingly simple goal of 5.7 million acres (2.3 million hectares) of newly declared wilderness in Utah (Shapiro 1998). Using '5.7' as a rallying cry, the group generated support throughout the United States and Canada, growing from zero to 20,000 donors in less than a decade. More importantly, the group was able, through wise and prudent communication strategies, to take an administrative problem and transform it into a broad public issue. The federal Bureau of Land Management was ordered in 1976 to study its Western land holdings and make recommendations to Congress for wilderness designation. In Utah, the BLM found 1.9 million acres (769,000 ha) worthy of this protection. Environmentalists disagreed and SUWA was established. The group's organizers decided to forego political lobbying and, instead, build grassroots support. They drew on Utah's sole metropolitan area, the Wasatch Front where two-thirds of Utahns live and then moved their campaign to the East and West coasts. The messages used by SUWA tapped into the beauty of the 'intricate canyons, arches, and vast expanses of slickrock' in Utah (Shapiro 1998, p. 265), as well as the learned legitimacy of scientists familiar with the ecology of this country. Attention also had to be given to the fears of residents of southern Utah, who felt they would be removed from their land. Throughout, SUWA successfully built a powerful coalition, by skillfully blending mass-mediated and interpersonal messages, taking alternately cooperative and confrontational stances, and creating an environmental issue that would not exist without their championing it.

SUWA and their beloved Utah wilderness illustrate numerous dimensions of issues. They are nurtured into wider consciousness. They call for a mix of messages to further the cause. They require attention on many levels – regulatory disputes, conflict management, public relations, legal repercussions, and health aspects. An issue cannot arise without confrontation and conflict, though these can be diminished. Opponents who are not adequately communicated with can erupt out of fear and deep disagreement. A complex situation can get unduly complicated. Managing messages is key to handling issues. And, understanding is a prerequisite to crafting successful messages.

4.4 Dissecting Issues

Analyzing issues is a lot like dissecting a specimen in a laboratory. You have to get beyond the surface to see what is going on inside. Different parts serve different functions. Perhaps the most difficult item to deal with is finding the core argument held by each side. Opposing groups many times have tremendous difficulty grasping what others are actually contending. Clarity of rhetoric is rare in environmental issues. Groups posture and position themselves. These stances shift constantly. Getting the upper hand in a battle of argumentation seems to require continual rearranging positions and an arsenal of rhetorical strategies. It is common to see two groups in a heated debate getting exasperated with each other because neither seems willing to see the other's viewpoint. This is often because the groups are locked into a confrontation over mis-aligned problems within their issue of interest. The groups are not arguing about the same viewpoint at the same

4.4 Dissecting Issues

time. Though each problem may be set within a general framework of an issue, they still may not be parallel. Arguing about them as if they are can only increase frustration.

The famous case of logging in the U.S. Pacific Northwest versus protecting the spotted owl clearly shows this situation. One group is arguing about jobs and lifestyles while the other is arguing about ecosystems for an endangered species. While both viewpoints are valid, it is necessary to realize that each viewpoint must be resolved independently.

Communicators need to be ever-vigilant about the multidimensional nature of issues. Confusion and clarity can both result when discussing issues. As groups declare their positions and discredit those of their opponents, watch for these characteristics within their perspectives:

- **Biased** Over-emphasizing a selective viewpoint demonstrates bias. 'Propaganda' is material espousing one and only one position. Even though information put forth by issue-interest groups is usually agenda-driven and meant to persuade you to adopt their point of view, it is essential to issue resolution to understand all perspectives in an issue if lasting resolution is ever to be achieved.
- **Simplified** Sometimes so much information about an issue is left out of an explanation that the message is misleading. Certainly, making a position easy-to-understand is laudable. Leaving out relevant details is not, however. Over-simplification is often a problem in risk communication scenarios, where crucial scientific details are omitted by message-makers who see shortcomings in their audience's scientific background instead of their own explanatory skills.
- **Personalized** When individual human dimensions of a wider environmental problem are highlighted, the issue is given a face. Such anecdotal evidence is powerful and persuasive. It tugs at your heartstrings. When used exclusively, though, personalization can cloud larger and more substantial aspects of an issue.
- **Sensationalized/Glamorized** Similar to being personalized, an non-local issue is sensationalized when the focus is on a single town, natural area, industry, or other entity. While the impact on the entity in the spotlight may be great, the connection to the wider society can be lost. When glamorized, an issue is championed by a celebrity figure. This may provide a lot of publicity, but the issue often plays second fiddle to the celebrity's persona.

People within groups sometimes erroneously think because they agree on an overall solution to an issue, their positions on answers to problems within the issue must be similar as well. Members of a group are more likely not to be in agreement on such details, even if they thought so on first brush. Values on which reactions to problems are based differ by individual. So, even within groups compromise will be necessary.

A vivid example of this occurred in the Sierra Club in 1998. At question was whether the Sierra Club should take an official policy for limiting immigration into the United States. Following the most extensive and vociferous internal debate in the club's history, members voted to maintain a neutral stance on this issue (Cone 1998). The campaign, which ended with a 60–40% vote, featured mass mailings, pithy sound-bites and accusations of racism. Even as this campaign created factions within the Sierra Club, there was no question that the nation's most powerful environmental group was united in its view that human overpopulation must be stemmed to reduce environmental degradation.

4.5 Value Descriptors

Values undergird all messages. They are tightly held by all within a communication system. Any lasting resolution of an issue will have to reveal and address the values of each group of stakeholders (Hungerford et al. 1988). Here are types of values within communication systems:

Aesthetic It is often said 'beauty is in the eye of the beholder.' What looks like a plain piece of scrubby grassland and briar patches to one person, may appear to another like a small desirable piece of wild area inhabited by numerous small forms of wildlife. A carpet of dandelions may be beautiful to one person and a scourge of weeds to another.

Cultural Many communities may have conservative approaches to doing something, because it suits them to do it that way and it has 'always been done that way.' Challenges to traditional methods of doing something usually meet with resistance.

Ecological People tend to resist even incremental changes in an area that is unique to them. Putting a road across a wooded hillside may create a barrier to wildlife.

Economic How might actions affect the economics of an issue? Who stands to gain and who stands to potentially lose? In the case of loggers losing their ability to harvest trees on national forest land, how are they to continue to make a living?

Educational What unique learning experiences can be gained from the issue and also from the process of managing the issue?

Egocentric Refers to a focus on self-satisfaction and personal fulfillment; a 'me' oriented value. What is in it for me? Even altruism has been shown to have egocentric dimensions.

Legal People may like or dislike legal values depending on how it affects them personally. If a law prevents hunters them from hunting out of season, they may challenge it. Yet another group wishing to stop hunting may support and even campaign to extend the non-hunting season.

Recreational How people spend their leisure time is important to them. If an issue affects their recreational activity or takes too much time away from it, then resistance will be encountered.

Spiritual/Religious A position on an issue may correspond to that of an organized religion, or to the more esoteric personal connection with something beyond the human world.

Social People come together for a variety of shared reasons such as empathy, feelings or status. Saving the rain forests of South America has drawn many people on other continents to join organizations to stop the clear-cutting of these forests. In these groups these people feel camaraderie and a feeling that they are doing something useful together.

4.6 Global vs. Regional vs. Local Issues

Another potent value descriptor and angle for analyzing an issue is its geographical coverage. The area affected by an issue will be a vital consideration in knowing who stakeholders are, as well as knowing what media are available to reach these stakeholders. Few issues are found in restricted ranges, single towns or small political units such as a county or riding. They are honestly local. More often, issues reach farther and have larger areas of effect, covering a region. Once effects are found across international boundaries, issues move toward being global. Watersheds are many times used to define the area of influence of an issue. Watersheds lend themselves to grading from local to regional, as they are nested, small ones inside larger and larger ones. Continents, hemispheres, climate zones, oceans, and airsheds also can be used to describe the geography of an environmental issue.

A curious finding by a series of researchers, environmental issues display strong distal effects. The newer and farther away an issue, the more likely people see easy and straightforward solutions to it. The distal effect runs counter to qualities of news needing to be proximate. In the early 1970s, Hungerford and Lemert (1973) noted a phenomenon they labeled 'Afghanistanism' in environmental newspaper reporting. They defined the effect as a presumed greater severity of environmental problems afflicting regions outside of a newspaper's home region. In Oregon newspapers, they found a concentration on issues outside circulation areas – as they put it, 'up the road a piece.' Roush and Fortner (1996) described another case of the distal effect, as zebra mussels invaded North America. Newspaper coverage of zebra mussels became less urgent and less likely to offer remedies between 1988 and 1993, while simultaneously the range of this invasive and damaging species of mollusk expanded rapidly. More recent and related studies hint that the distal effect may reverse, at least for Americans, once an issue goes global. Konisky et al. (2007) found less public support for government action on global climate change than for regional resource issues and local pollution issues. Kellstedt et al. (2008) similarly found consumption of more information on global warming correlated with lower concern and a shrinking sense of personal responsibility on the issue. Soberingly, their results concur with 'Green at Fifteen?' (2009), where the most advanced students were also the most pessimistic.

In putting boundaries on the geographic scope of environmental issues, communicators may wish to err on the side of casting their net a bit wider, rather than forcing narrower boundaries. Allowing a larger range for an issue can help inoculate messages against any distal effect. As a bonus, a wider definition of place might also assist in connecting to consumers of messages within the placeless-ness of the Internet.

4.7 Framing and Framing Anew

A concept useful in describing and analyzing group positions, framing refers to the perspective from which an issue is viewed. To frame is to select certain dimensions of the issue at hand and grant them more salience than others (Entman 1993). Messages are framed by the way their component bits of information are included, connected and arranged. Frames define issues, diagnose causes of problems, and suggest answers and solutions. Conversely, framing always involves exclusion of portions of an issue. Frames are based on values that drive the selection process and the measuring of salience. The term 'frame' is itself a metaphor, helpful in generating an understanding of an intangible mental process. Like a border on a painting or photograph, a mental frame sets off the subject from the rest of the field of view. Frames put boundaries on reality, tend to be self-reinforcing, and are difficult – but not impossible – to overcome (Entman 1993).

One's initial task in revealing a frame is to find out who controls the composition and release of information. Historically and currently, information-providers have tremendous power, and power is the prime factor in pushing one's own agenda over somebody else's. Therefore, information and the way it is framed exerts control over the audience. Audiences have been shown to take cues from frames, learning what aspects within issues to pay attention to and what connections exist between problems within an issue.

Environmental communicators should look for indications of frames as they gather information about issues. This need not be a strictly academic exercise, as long as one considers that frames have power to impose meaning. If one can find signs of the particular frame in place, one can begin to construct alternative frames. Changing frames is possible and useful. Done effectively, new frames can build dialogue and result in more harmonious relationships among groups (Ryan 1991).

Reconstruction of frame is a typical tactic used in environmental communication. In using this re-framing tactic, you change the original point of contention in order to obscure an opponent's argument, gain support from a broader audience, or align your own message with some value of the target audience.

In establishing a regional outreach program, the Idaho Museum of Natural History conducted audience analysis to better understand the knowledge, interests and attitudes of the people of the U.S. state of Idaho (Sommer 1999). This front-end evaluation showed the museum that traveling exhibits, educational videos, and traveling trunks they wished to develop and disseminate needed to reflect the values of rural Westerners if they were to facilitate learning about biodiversity. They decided to do this by using language that was common to and acceptable by the audience, even if it was unconventional for natural history museum staff. The outreach program was built on a teaching metaphor, 'the economy of nature.' This metaphor is used instead of 'ecosystem' which was found to be potentially threatening. Likewise, instead of 'environment' and 'evolution' the museum used 'the great outdoors' and 'natural selection.' The political baggage carried by the conventional terms was, thus, avoided. The reworking of language amounted to a realignment of the frame embedded in the museum's educational messages with the values of its intended audience. Curators and educators wisely adopted an alternative frame that would facilitate learning by their audience, rather than generate noise in the form of confusion and anger.

A pair of cautions is in order. First, lapsing into propaganda is easier than it should be when constructing an alternative frame. Propaganda is not always the telling of lies. It can also be presenting only selected truths to support your position, while completely neglecting other viewpoints. Second, when issues are redefined it is common to retain presumptions from before. These should be examined and probably changed as well, however. This keeps groups from assuming they are still arguing the same points as before and getting into the uncomfortable and unproductive situation of 'talking past each other.'

Properly understanding issues is the first, vital step in attempting to manage them. This chapter has dealt with the organization of information as it pertains to revealing the complexities of environmental issues and the values of the groups staking and defending positions on such issues. Understanding an issue clearly makes the prospect of getting all participants talking about the same topic at the same time achievable.

4.8 Case Study: Environmental Issue Analysis. Ohio Beverage Container Deposit Legislation

Reframing serves a valuable communication function by directing receivers' attention to points of interest within an issue desired by the communicator. A danger, however, is that reframing can cloud an issue as well as clarify. Ohio's Beverage Container Deposit (BCD) legislation issue shows how reframing can be used to obfuscate, and negatively affect issue resolution. Starting around 1980 in the American state of Ohio, there were almost annual drives to pass a BCD law. A group called Citizens for the Environment wanted a deposit placed on beverage containers to prevent them from being discarded after use. Such legislation would provide a financial incentive to bottling manufacturers to pick up containers for recycling from consumers. Opponents of BCD do not want to place such a burden on industry. In 1994, the beverage industry redefined the issue completely. They offered taxpayers an alternative frame by declaring the penny-per-bottle tax was a tax on food. Food taxes are both unpopular and unconstitutional in the state of Ohio. Industry proponents placed a referendum on the November 1994 ballot and launched a campaign that convinced Ohioans to repeal this tax. The bottling industry was able to concentrate the public focus on food tax and not litter reduction and recycling. Voters repealed the tax, thereby taking $64 million out of the state's annual budget and giving it back to the bottling industry. To have gotten their desired outcome Citizens for the Environment would have had to, as George Lakoff (2004) says, confront the obfuscation quickly and defend the frame of recycling.

Credit: Ohio Beverage Container Deposit Legislation, Clipart 'Bottle' Word 2009 Office.

References and Further Reading

Bryant B, Bezdek R, Ferris D, Kadri J (1995) Environmental justice: issues, policies, and solutions. Island Press, Washington, DC

Buckingham S, Turner M (2008) Understanding environmental issues. Sage, London

Cone M (1998) Sierra Club to remain neutral on immigration. Los Angeles Times, April 26, 1998

Entman RM (1993) Framing: toward clarification of a fractured paradigm. J Commun 43(4):51–58

Easton TA (2008) Taking sides: clashing views on controversial environmental issues, 13th edn. McGraw-Hill/Dushkin, Iowa

Green at fifteen? How 15-year-olds perform in environmental science and geosciences in PISA 2006 (2009) Programme for international student assessment, Organisation for Economic Co-operation and Development, Paris

Hungerford SE, Lemert JB (1973) Covering the environment: a new 'Afghanistanism'? Journal Q 50:475–481, 508

Hungerford L, Peyton R, Volk T (1988) Investigating and evaluating environmental issues and actions: skill development modules. Stipes, Champaign, IL

Kellstedt PM, Zahran S, Vedlitz A (2008) Personal efficacy, the information environment, and attitudes toward global warming and climate change in the United States. Risk Anal 28(1):113–126

Konisky D, Milyo J, Richardson L (2007) Environmental policy attitudes, political trust, and geographical scale. Paper presented at the annual conference of the Western Political Science Association, San Diego, CA

Lakoff G (2004) Don't think of an elephant! Know your values and frame the debate. Chelsea Green, White River Junction, VT

McConnell RL, Abel DC (2001) Environmental issues: measuring, analyzing, and evaluating, 2nd edn. Prentice Hall, Upper Saddle River, NJ

McConnell RL, Abel DC (2007) Environmental issues: an introduction to sustainability, 3rd edn. Prentice Hall, Upper Saddle River, NJ

Ramsey JM, Hungerford HR, Volk T (1989) A technique for analyzing environmental issues. J Environ Educ 21(1):26–30

Roush D, Fortner R (1996) Newspaper coverage of zebra mussels in North America: a case of 'Afghanistanism'? Electron Green J 1(5):article 3

Ryan C (1991) Prime time activism: media strategies for grassroots organizing. South End Press, Boston, MA

Shapiro DW (1998) Lessons from a successful grassroots environmental campaign: the southern Utah wilderness alliance. In: Senecah SL (ed) Proceedings of the fourth biennial conference on communication and environment Syracuse. State University of New York College of Environmental Science and Forestry, NY, pp 265–270

Sommer SA (1999) Opening doors of communication. Presentation at the annual conference of the American association of museums, Cleveland, OH (April 1999)

West B, Sandman PM, Greenberg M (2003) Reporter's environmental handbook, 3rd edn. Rutgers University Press, New Brunswick, NJ

Part II
Communication Planning

Chapter 5
Planning Environmental Communications

5.1 Introduction

Messages are most likely to be effective when they are part and parcel of planned campaigns. The process of communications planning formulates your campaign's goals and objectives, analyzes your intended audience, marshals available resources, and sets a schedule for its implementation. A plan's purpose is to harness and focus the power of the resulting communications system, to make it efficient and effective. This chapter presents an outline for writing plans for environmental communications campaigns.

As with any planning, it is handy to remember U.S. President Dwight D. Eisenhower's 1957 admonition: 'Planning is everything; the plan is nothing.' Planning is a methodical approach to a process. Once completed, the plan may need to be revised during implementation. As conditions change, so should your plan. Though it is important to be as complete and detailed as possible, flexibility in the execution of your plan is advised if you find audiences reacting differently than anticipated. Flexible plans contain alternative actions for the most likely contingencies, and are instilled with the realization that the unpredictable can, from time to time, happen.

A planning process is diagrammed below. Each of its parts is detailed in the remainder of this chapter and elsewhere. Keep in mind that these parts are steps in a process. After detailing the process, we present a generic format for the plan document. Even though what we layout here looks like a linear procedure, in reality planning is iterative and often roundabout.

5.2 A Process for Planning Campaigns

The schematic below focuses on discrete steps in planning and executing a campaign. Each step needs to be considered carefully, though not necessarily sequentially. In reality, these steps blend into each other, and may vary in sequence (Fig. 5.1).

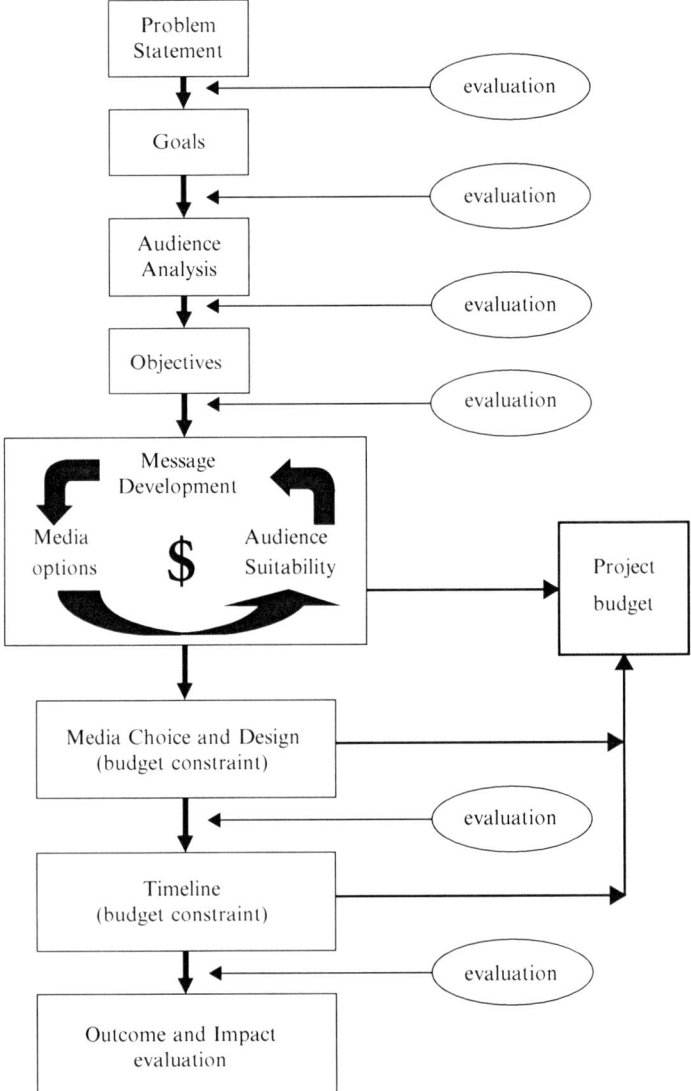

Fig. 5.1 Communication Planning Model. Original graphic, Richard Jurin

5.2.1 Problem Statement

Communication campaigns are launched in response to specific issues, problems and needs. Issues are wonderfully multidimensional, so planning must be too. A planning process begins with a thorough analysis of the subject to be addressed. During issue analysis, record as much detail using descriptors about each of the issues components as you can. Try to concisely define the frame. After careful consideration and exploration of the aspects of the issue, a short but complete problem statement should be developed. Your problem statement guides the rest of the planning process. Research has shown that how environmental problems are defined affects how solutions to them are developed (Clark 2005). To understand the issue from other points of view, outside contributions can be solicited when developing a problem statement. Since issues have multiple problems within them, more than one problem statement may be necessary to adequately address an issue through a campaign.

After drafting a satisfactory problem statement, it should undergo evaluation to determine whether your statement adequately covers your position on the issue and is understandable by the intended audience. To accomplish this, you might convene a focus group to check for consensus on the adequacy of the problem statement. You may even have to collect data via a needs analysis to identify just what is required. Regardless of the method you use, care should be taken to consider a range of opinions.

5.2.2 Goals

Once you are satisfied with your problem statement, develop goals for the communication campaign. Goals represent the ultimate aims of the project, and are best when they reflect long-term, lasting outcomes. In general, goals point to a resolution of the problems identified by the problem statement. Inclusive, open wording of goal statements allows more creativity in developing your campaign; specificity should be reserved for objectives as discussed below. Goals can be thought of as *qualitative statements of desired end-states.*

5.2.3 Audience Analysis

At this stage in the process, communication planners work on identifying and gathering information about the specific audiences of the campaign. An audience can be conceptualized as a group of people who can be reasonably expected to react in similar ways to a message. They are those to whom your messages will be delivered.

Audience analysis involves repeated bouts of defining and information gathering; it is iterative. As planners learn more about potential audiences, they refine their definitions of them. Well-done audience analysis makes other planning activities easier. So, it is important not to rush through this critical step in the planning process.

5.2.4 Objectives

Objectives are specific, measurable outcomes to be achieved by your messages. Objectives should be expressed in a way that makes it clear when they have been achieved. In addition, objectives should be stated as impacts outside of direct control of the communicator, not as actions to be taken. This makes them distinct from the chronological steps of the timeline, directly under the communicator's control. Objectives are usually expressed in terms of a target audience taking some measurable action by some date. Examples of objectives are:

- Seventy percent of students at Smith Elementary School will recycle their soda cans by June.
- All park visitors will be aware of new trail regulations after January 1.

Objectives that are too vague for measurement, involve poorly defined audiences, or are expressed in terms of activities within the control of the communicator are not supportive of good planning. Examples of *poorly conceived* objectives include:

- All members of the state chapter will be more environmental. ('More environmental' is too vaguely worded for measurement.)
- The general public will support the timber sale scheduled for June 30. (The audience 'general public' is too poorly defined for measurement.)
- The communication team will issue a news release by December 1. (This is completely within the control of the team; it is not based on the audience.)

The definition of objectives is a critical step in communication planning. In short, objectives can be thought of as *quantitative statements of the campaign's benchmarks and milestones.*

5.2.5 Message Development/Media Options/Audience Suitability

After audiences are identified and objectives set, communication planners must formulate a message (or messages) to achieve their objectives. Simultaneously, media options, otherwise known as communication channels, should be researched for their abilities to deliver your messages. Combinations of messages and media are considered with your audience in mind. For example, simple messages designed to raise awareness for large publics might be planned for use with television or billboards, as long as you can confirm that the target audience use these particular channels. Narrower and more diffuse audiences require more selective media choices. Like so many cases in environmental communications, media placement is an iterative process, where continual review of your messages, media, and audience is imperative.

In order to move past this step of the planning process, a campaign cost limits will need to be established. Some media options may be beyond the financial resources of a project. For example, paid television advertisements will probably be beyond the monetary limits of small nonprofit groups. A new media alternative, targeted Internet advertising has opened new doors for environmental communicators with big ideas but small budgets. Though a detailed budget is not needed at this stage, simple aggregate spending limits are required for the plan to develop.

5.2.6 Media Choice and Design

After media options are identified, specific choices must be made. Given overall budget constraints, communication planners must decide which media options most likely will be effective in delivering the message and achieving the plan's objectives. Preliminary design should occur at this point, to guide implementation and budgeting. For example, if magazine advertisements are to be used, preliminary design work would include advertisement size and frequency, with mock layouts needed to continue the planning process. If a web site is part of the plan, there needs to also be plans for driving traffic to the site.

5.2.7 Timeline

When media have been chosen and preliminary designing completed, communication planners schedule the remaining steps needed to implement the plan. These steps should include specific actions to be undertaken, dates steps will be started and completed, and an assignment of responsibility for completing each step. This phase is completed when the communicator has developed an overall project timeline. A well-formulated timeline allows the communicator to carefully monitor a project's progress and to quickly identify bottlenecks, should they arise.

5.2.8 Front-End Evaluation

Evaluation is a systematic examination of the development of a campaign against standards of one sort or another, as well as a measure of the campaign's effectiveness and progress toward goals and objectives. Depending on when during the campaign's creation and roll-out evaluation takes place, it can be classified as front-end (early in the game), formative (in the middle), or summative (toward the end). Front-end evaluation happens early and informs design of the campaign. Samples of front-end evaluation include identification of audiences; literature reviews; small-scale tests of sample messages and media; and expert review of problem statement, goals and objectives.

5.2.9 Formative Evaluation

Formative evaluation takes place within the planning process, during the formation of your campaign. Think of formative evaluation as fast-looping feedback, to assess whether a project is meeting its objectives. In communication planning, formative evaluation is an important tool to verify assumptions and decisions made during the process. Evaluation makes it possible for errors in assumptions or poor decisions to be corrected without wasted resources.

After a problem statement is formulated, planners can use formative evaluation to be sure the statement accurately reflects the problem as seen by the sponsor. Likewise, goals can be verified using consensus techniques with samples from stakeholder groups. Formative evaluation can be used to confirm that audiences are correctly identified, and to test messages and media with specific audiences. Rarely are complicated and expensive communication projects launched without first testing their components for effectiveness.

5.2.10 Summative Evaluation

Communication planning is incomplete and risky without evaluation built-in throughout a campaign. Environmental communication plans should always include a description of how the project will be evaluated. Specifically, planners should identify specific means to be used to analyze whether the project has successfully met its objectives. All too often, communication efforts simply assume that the objectives will be met and never actually measure successes or failures. Most funding entities now insist evaluation be conducted, however. This creates accountability for resources being used in the communication campaign. While well-conceived objectives will ease the evaluation effort, specific evaluation techniques should be identified well before you need to call on them to see if objectives have been met.

5.2.11 Project Budget

Sound financial management is part of all good planning processes. A detailed budget should include the cost of materials, labor, purchased services, and overhead expenses for the project. One method for developing a budget is to review the project timeline and identify the costs of every step on it. Many planners underestimate the real costs of communication projects, and are later forced to prematurely end or scale-back their effort, and thus not achieve their objectives. Detailed, accurate budgets greatly increase the professionalism of communication plans. While budgets are mostly about financial resources, they also inform the allocation of other resources such as human capital and time.

5.3 An Outline for Writing a Communication Plan

Each communication plan must take its own form. That said, the following format suggestions can be effectively used for compiling a communication plan. Such a plan should be committed a proper form, be it paper or digital, and professionally presented. Amendments should be added to include additional material as necessary. As with all written communication, clarity, completeness, and brevity are important.

- **Abstract** A one-page summary of the plan should be included at the beginning of the document. This section is also known as the executive summary.
- **Introduction/Background** The problem statement and a concise history of the subject being addressed, at the least, should be included in the plan. Planners cannot assume that all users of the plan understand the issues being addressed.
- **Goals and Objectives** The campaign's goals and objectives, as defined in the planning process, are included at this point. Remember that objectives should be specific, measurable, and stated as effects on the audience.
- **Target Audiences** This section describes the groups at the receiving end of the campaign, including specific information about them and inferences made in designing and delivering messages to them.
- **Implementation** This section should present the message(s), media choices with justifications, and the details of the plan implementation. Included in this section are the project timeline and a detailed budget. Contingencies for revamping the campaign during implementation may also be included here. This section should be as detailed as possible, so an individual who did not assist in preparation of the plan could still use it for implementation.
- **Evaluation** This section should tie closely to the objectives. Evaluation methods should be clearly and succinctly described, along with criteria for determining the success or failure of the project. Evaluations may include numerous types of data collection or just one. The determination depends on what the plan and objectives warrant.
- **Budget** This section should itemize and detail all the costs of the plan. This helps readers to see all the expenses involved and assists decision-making when adjustments are needed.

Planning is a prerequisite of success. It helps identify what needs to be done and the best ways to achieve the required outcome. A good plan uses resources wisely and efficiently, helps in avoiding unexpected problems, and moves an issue toward resolution.

5.4 Case Study: Environmental Communication Planning

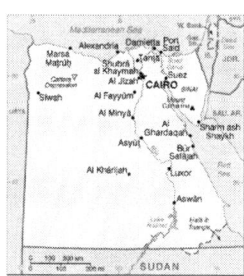

Egypt's national policies for facing its environmental challenges include an environmental communications plan. The strategy, covering 2005–2010, is extraordinary as a guide for an entire government in a developing country. The document runs 61 pages, a credit to the author's ability to present a fleshed-out plan without it being bloated. Compare the names of components we list with equivalents from Egypt's plan: *Problem statement* 'Strategy for environmental communication, why?' (Hassan et al. 2005).

Goal 'General Objective – to provide communication support for the various environmental efforts aiming at achieving sustainable development, with the final goal of improving citizen's quality of life and achieving welfare for future generations.'

Objectives 'Communication Objectives,' e.g., 'Putting environmental issues on top of public priorities and increasing its space in social communication among social groups.'

Target Audiences 'Target Group Priorities,' e.g., 'Professional public, including farmers; fishermen; managers of medium-size enterprises in the fields of agriculture, industry, tourism; owners of small-size enterprises.

Messages 'Cognitive and behavioral messages,' e.g., 'Environment protection is a social responsibility on every institution, organization and individual in Egypt.'

'Protecting and preserving the environment is an ancient Egyptian behavior encouraged by religions and by Egyptians' culture throughout years.'

'A healthy environment is a source of abundance of good for you and your family.'

Development included heavy front-end and formative evaluation. Implementation plans are detailed through extensive matrices, showing target audiences, sample messages and media, and sub-objectives including quantified desired behavior changes. The plan could be strengthened with the inclusion of a one-page abstract and a detailed budget. Both these components would assist in fostering buy-in from personnel throughout the Egyptian Environmental Affairs Agency who are tasked with implementing this ground-breaking plan.

Credit: Egypt map. The World Factbook.

References and Further Reading

Allen J (2009) Event planning: the ultimate guide to successful meetings, corporate events, fundraising galas, conferences, conventions, and other special functions, 2nd edn. Wiley, Brisbane

Allison M, Kaye J (2003) Strategic planning for nonprofit organizations, 2nd edn. Wiley, New York

Bryson JM (2004) Strategic planning for public and nonprofit organizations: a guide to strengthening and sustaining organizational achievement, 3rd edn. Jossey-Bass, San Francisco

Cassidy A (2005) Practical guide to information systems strategic planning, 2nd edn. Auerbach, Boston, MA

Clark TW (2005) Averting extinction: reconstructing endangered species recovery. Yale University Press, New Haven, CT

Ferguson SD (1999) Communication planning: an integrated approach. Sage, Newbury Park, CA

Goodrich W, Sissors J (2001) Media planning workbook, 5th edn. McGraw-Hill, New York

Hassan H, Nasr E, Al Gazar N (2005) National strategy for environmental communication. Egyptian Environmental Affairs Agency, Ministry of State for Environmental Affairs, Cairo

Smith R (2009) Strategic planning for public relations, 3rd edn. Routledge, Buffalo

Walker TJ, Todtfeld J (2008) Media training A-Z, 5th edn. Media Training Worldwide, New York

Wilson LJ, Ogden JD (2008) Strategic communications planning: for effective public relations and marketing, 5th edn. Kendall/Hunt, Dubuque, Iowa

Chapter 6
Analyzing Your Audience

6.1 Introduction

Selling a product requires a thorough knowledge of the potential buyers. Marketers acquire and apply such knowledge about consumers. Likewise, environmental communicators anticipate particular reactions to their messages by audiences. Sending messages that produce desired effects requires a thorough knowledge of the groups with whom you will communicate. Audience analysis comes early in the communication planning process for many reasons, because appreciating and catering to the attitudes and opinions of the groups your messages reach is crucial to the success of a campaign.

A target audience is any group for which a message is specifically developed and intentionally focused. An intended audience is one that the communicator expects to react to a message. This is not everyone who might come into contact with the message. The more you know about your intended audience, the more likely the message is to be received and acted on in accordance with your campaign's goals. By action, we mean anything from becoming aware of a situation to permanently modifying a behavior within the audience. As you might suspect, making an audience aware of something is a lot easier than changing their behaviors. In this chapter we will deal with several aspects of analyzing audiences to promote an understanding of why people may or may not respond to your messages.

Learning about those who are targeted by your messages concentrates your campaign, keeping you from trying to reach too broad an audience. Audiences can be assessed by answering such questions as:

- What is the message? Why send it?
- Who is (are) the audience(s) that need to be addressed?
- Why communicate with them? What makes them important to the success of your campaign?
- What makes them special and how can you customize your messages to meet their needs?
- What do you want them to do with this information? What kind of reaction do you expect?

- How will you know that they understood your message?
- What was their reaction to your message?

Keep in mind there is no such thing as a monolithic group called 'the general public.' Even though this term is heard often, it is far too imprecise to serve a useful purpose to environmental communicators. Using this term shows audience analysis is missing or inadequate. Even if you have the aspirations and budget to try to reach billions of people, they still are not everyone. No message yet has reached all seven billion humans on earth. Those with the widest reach – logos of such mega-corporations as Coca-Cola, Nike, and General Electric – may be seen by most, but nowhere near, all of us.

For environmental messages, there are many possible audiences. Each may be classified into a common group based on their shared characteristics, interests and demographics. These are only indicators of your audience, however. Moreover, a group that accurately constitutes an audience for one type of message may not hold together for a different message. Audiences are fluid and ever-changing. Think of them like groups of people using an elevator: every trip up and down has a different makeup. Thorough research into your target audience is essential to maximize success with your communication efforts.

Recall our communication models. There should be a need to communicate that requires you to send a message to specific receivers. To reach them you'll have to overcome noise that interferes with the system. Encoding the message correctly will help reduce noise so that your target audience receives and decodes your messages as intended. Correctly analyzing your audiences is one major way of overcoming barriers to communication and ensuring that your message gets through.

Audience analysis usually involves talking to or surveying a select group of people that are similar to the wider audience you wish to reach. Such investigations are part of the planning process that goes hand-in-hand with having clear goals supported by measurable objectives and well thought-out evaluations. If you cannot get firsthand information about your audience, then reviewing other situations that resemble yours can help. Interviewing of key people who know about your audience is also beneficial in helping you develop a profile of your audience's backgrounds. The more you understand your audience the more likely your message will 'hit home.' What follows in this chapter are some ideas to help in analyzing audiences and understanding how messages are received and acted upon, or – as is so often the case – ignored.

6.2 Internals Versus Externals

One of the most basic divisions in audiences is between internal and external publics. Internals are people directly involved with an organization. They are on the inside, as employees, members etc. They can be expected to identify with the goals and missions of the organization producing the message. They also tend to have a vested interest in the outcome of communication campaigns. Externals, on the other hand, have no vested interest in the organization and a message aimed at them must emphasize the 'so what' question (Fig. 6.1).

INTERNALS	EXTERNALS
Non-Governmental organizations	
Full time employees Seasonal staff Retirees Volunteers Members Board of directors Corporate donors	Community groups Civic associations Consumers Recreationists Business community
Governmental organizations	
Full time employees Seasonal staff Retirees Volunteers Contractors Advisory boards Legislators	Visitors Neighboring Communities Special interest groups Business community Non-elected leaders

Fig. 6.1 Internal versus external audiences. Original graphic, Richard Jurin

6.3 Population Segmentation

Another way of viewing groups of people and how they will respond to new ideas and actions is through population segmentation. For this type of audience analysis, a population is divided into groups, for which profiles are drawn explaining expectations for responds to particular messages. Actually knowing your audience is critical to success.

6.3.1 Adoptions of New Ideas

Adopter distributions in the American population follow a bell-shaped curve over time (Rogers and Shoemaker 1971; Rogers 1995). This segmentation of the American population describes how likely they are to adopt new ideas and technologies. A communicator will need to assess how 'average' their target population is and to which categories they are likely to fall into. The following categories give useful insights into a population:

- **Innovators** (2.5% of the population) – These venturesome few are the vanguard for new ideas and behaviors. They tend to initiate 'new things' and take social risks in their adoption of new ideas. They are most often wealthy and socially established with large-ranging influence. They are also information seekers getting their information from primary sources, or may even be sources of new information. While they are trendsetters, they are rarely used as advice-givers. Rather, others look to them and imitate.

- **Early Adopters** (13.5%) – This segment is respected by others as being continually innovative and ready to try new things. They include both formal and informal leaders of organizations with social influence. They tend to be young and knowledgeable users of in-depth and specialized sources of information. Both innovators and early adopters prefer hard evidence and factual accounts over unverified anecdotes. They also are the kind of person compelled to have the newest 'whiz-bang' gadget before everyone else. They are trendsetters and also recommenders.
- **Early Majority** (34%) – This large segment is deliberate in their decision-making about new ideas and trends. They include many informal 'quiet' leaders with influence within smaller social groups and small communities. They tend to be young adults to middle-aged with modest financial means. Though they are still opinion leaders, they are viewed as discriminating by others. They do not accept, or even test, every new idea they come into contact with. Their information sources are commonly family, friends and mass media, in that order.
- **Late Majority** (34%) – A skeptical group, the late majority usually adopts only established ideas. They tend to be older with moderate education and 'keep to themselves.' They have conservative lifestyles and learn much of their information from acquaintances, friends, and family, rather than mass media.
- **Laggards** (16%) – These people define the term 'traditional.' They adopt ideas reluctantly and only after they are unable to avoid change. They perceive great social risk in new things. They are often older people and those with the least education, lowest incomes and lowest social status. Friends and family of the same social station are their primary sources of information. Sometimes they ignore change agents and rebel if forced to try and adopt something with which they do not identify.

6.3.2 Support of Pro-environmental Issues

Roper (1990) categorized the American public into five major groups based on their environmentally responsible behavior, awareness and consumerism. Remarkably, political affiliation was not a factor in establishing these segments. And, a sense of external control of affairs became more predominant moving down the list.

- **True-Blue Greens** (11% of the population) – Like the Innovators, True-Blue Greens tend to be affluent and respected members of society. Interestingly, they are comprised two-thirds of women. They have the highest incomes, are more educated, and tend to be innovators within their communities. Their behaviors tend to be consistent with their strong concerns about the environment.
- **Greenback-Greens** (11%) – This group can be described as affluent with high incomes and substantial education, though below that of True-Blue Greens. They also demonstrate less environmentally responsible behavior. They are, however, large supporters of environmental organizations, through monetary donations. They will pay slightly more for environmentally friendly products, but do not usually become involved in environmental activism.

- **Sprouts** (26%) – As a swing group, they can be pro-environmental or anti-environmental on any particular issue. Their concerns wax and wane, depending on the individual saliency of the issue at hand. Overall, they are near-average in income and education. They show the characteristics of 'typical middle America.' While they have green tendencies, they show no enduring pro-environmental behavior. They are rarely willing to pay more for green products and almost never become activists.
- **Grousers** (24%) – Generally anti-environmental, lower income and less educated, individuals in this segment consistently rationalize their lack of environmental behavior by offering excuses. They also criticize the poor performance of others.
- **Basic Browns** (28%) – This group holds few opinions on environmental issues. Their behavior is not environmentally friendly. Their incomes are low. They tend to be poorly educated and are predominately male. Most work as unskilled labor. Of all five groups here, Basic Browns have the lowest levels of action and actually resist efforts for environmental improvements as an attack on people like them.

6.3.3 *Fragmentation, Selectivity and Loyalty*

Classic studies segmenting the American public as adopters (Rogers 1995) and by how they behave environmentally (Roper 1990) remain relevant by shedding light on relatively stable factors, basing their findings on large samples, and by peer-acknowledgement through wide citation. Few audience analyses achieve such fame, nor are few designed to achieve such heights. Rather, a typical audience analysis is an internal, proprietary tool to equip communicators representing an agency or organization as they make a campaign.

Machining the right tool for the audience-characterizing job at hand may be getting more difficult. Through the 1990s and accelerating into the twenty-first century, ever-heavier Internet use and converging media trends has broken audiences apart and reshuffled them repeatedly. Communications planners are attempting to describe audiences that are more fractured, more disperse, more selective, less geographically dependent, and less loyal, than in decades past.

Fragmentation has been shown empirically. In Europe, Dutch use of television and publications has become less predictable, based on loyalty to any one form within the media (van Rees and van Eijck 2001). The only trend evident between 1975 and 1995 across all Dutch readers and viewers was less reading. In the United States, audiences for television has gone from more than 90% concentration in three broadcast networks in 1977 to a panoply of more than 300 available networks, few of which ever score a double-digit share of the viewing audience (Webster 2005). He notes fragmentation may fuel polarization, too, as the average household receives more than 100 channels but uses fewer than 15. Rather than providing a central forum for national issues, television now is a cornucopia of distinctive networks catering to selective and narrowly defined slices of the viewing public. TV viewing segments now fall between small-and-loyal and small-but-disloyal (Webster 2005). Tewksbury (2005) followed news seekers as they left television for online sources. He found media users making individual decisions to hone in on particular topics on particular online sources – or as he wrote, 'ample signs of outlet specialization.'

Audiences are fragmenting because individuals are more selective in their media diets and have more to choose from across information sources. More so than in the past, audience analyses are only temporarily valid. Segments are mostly narrower and less persistent, so an audience analysis performed for one campaign will not likely stay fresh long enough to use for a subsequent campaign.

6.4 Adopting New Ideas

Through adopter categories, populations were segmented based on how they adopt ideas and technology. In reality, the process for populations is a more complex, however, since different people move through the adoption sequence at different rates. Our culture is built on ideas that have been widely accepted. Thus, some ideas become ideals and are called cultural norms. For example, we reject killing of others as wrong in our society. 'Thou shalt not kill' has been upheld as a norm for millennia.

Ideas flow through societal discourse, either gaining acceptance or being rejected. The adoption sequence modeled here functions in a similar fashion as the knowledge filter. People develop acceptance to an idea that the sender has to give to the audience. It is not instant acceptance, and indeed, the adage 'you can lead a horse to water but you can't make it drink' is applicable here in most cases. It takes a lot of effort from the communicator to move an idea through an audience from awareness to acceptance, even when you have a receptive audience. As an environmental communicator, you seek to interject ideas into the discourse of your audience.

The adoption sequence theory (Lionberger 1960; Rogers 1995) posits that ideas go through a five-step process before the behavior of a large segment of the population can be affected. As an idea moves from one step to another through exposure to more people, the amount of effort by a communicator to maintain the process usually increases. Sources of information used also shift as people begin processing information for deciding whether to accept or reject an idea. Ideas may first come to awareness through mass media, usually via Internet programming, television or popular publications. But adoption is more likely to driven by interpersonal contacts with respected and trusted friends. Factors affecting the adoption sequence are many and varied, and a wise environmental communicator will develop their plans based on sound evaluation and knowledge of the target audiences. Let's take a look at the steps in the adoption sequence (Fig. 6.2):

6.4 Adopting New Ideas

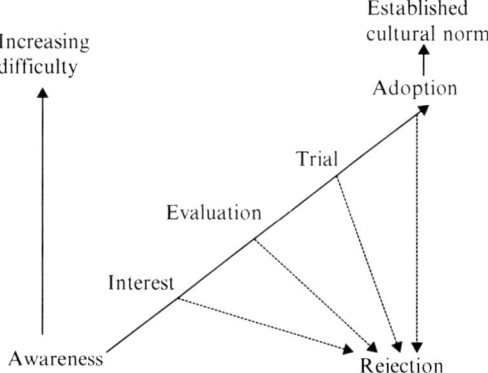

Fig. 6.2 The adoption sequence. Original graphic, Richard Jurin

6.4.1 Awareness

This is the easiest step for communicators to accomplish because it involves just exposing the audience to an idea. Get a shred of attention and you can claim to have generated awareness. Most often, this is accomplished through mass media, new media or a convergence of the two forms. The influence of friends and relatives is a secondary source of exposure, and tends to be more expeditious, rather than scheduled. Awareness does not necessary involve any changes in a person's attitudes and opinions. It also does not absolutely lead to retention of an idea or contemplation of its implications. Nevertheless, simply getting new information on a public's psychological radar screen is required, if further steps of the adoption sequence are to be attained.

6.4.2 Interest

Interest occurs when an idea resonates with an audience. To be interested the audience members need to be shown the salience and relevance of the idea. For this reason, the process becomes a little more involved at this stage. Usually, messages work to reveal to an audience how the new idea affects them in some specific way, either negatively or positively. Mass, new and combo media are still the primary modes of communication at this stage. Still, a communicator is working on knowledge acquisition within the audience. Attitudes and opinions have not yet been affected.

6.4.3 Evaluation

This is a critical stage, for here the purpose of messages changes from knowledge transmission to persuasion. Mass media are now less effective as people begin to consciously consider options presented within an idea. Advice from experts and other trusted opinion leaders is sought. Therefore, having credible sources for audience members to interact with becomes a crucial factor in maintaining an audience's progress up the sequence. People begin here to examine their own attitudes and opinions.

6.4.4 Trial

As individuals begin trying out the idea's upshots, they look for positive reinforcement from others. If attitudes and opinions have been called into question, alternative behaviors will be tested. Experts may need to increase personal interaction with the primary audience, or develop messages for secondary audiences that can be influenced by the primary audience. Acceptance is made more likely if the individual has a positive connection with the idea or behavior. The idea or behavior may be rejected at this level if benefits do not fit within an individual's belief structure.

6.4.5 Adoption

After favorable trials, an idea and its attendant behaviors can be adopted. An idea can still be rejected, even at this stage, indeed they often are. Adoption many times is impermanent. Since a communicator will have a lot invested at this stage, outcome and impact evaluation may be a sound investment to reveal what was unacceptable to the audience. This assessment will allow future efforts to be modified to meet goals and objectives. If lasting adoption is to take place, it is important periodically provide positive reinforcement of the adopted idea. Continuation of messages prevents the audience from regressing to their old positions and galvanizes their new adoption. All in all, it takes some time for new behaviors and ideas to become established as cultural norms.

Moving through the levels, cognitive functions become more and more sophisticated. Awareness and interest are considered knowledge functions, a matter of committing to memory. Evaluation is considered a persuasion function, whereas trial and adoption are decision functions (Solo and Rogers 1972). Knowing these functions helps the communicator focus a message for a specific purpose when viewed as part of the adoption steps.

6.5 Beliefs, Values, Attitudes, Worldviews, and Opinions

In referring to how people think and act, the words 'beliefs,' 'values,' 'attitudes,' 'worldviews' and 'opinions' are commonly used. As one delves deeper into communicating about the environment, these terms take on special meanings.

These meanings and their implications are important to state, so that later discussions about them can grow from a common understanding. Environmental communicators need to make many assumptions about their target audiences. Most of these assumptions involve a target group's shared beliefs, values, attitudes, worldviews and opinions. The complex created by one's beliefs, values, and attitudes can be referred to as their 'worldview.' Those with similar worldviews are predisposed to act in similar ways. The best course a communicator can take in considering the reactions generated by their messages is to really know their audiences and to understand that beliefs, values and attitudes are culturally derived.

6.5.1 Belief

A belief is a basic, deeply held ideal upon which people intellectually make sense of information. Beliefs structure what we think about other people, objects, places and issues. Many times they can be expressed as simple declarative statements. Examples of common beliefs are: 'Knowledge is good,' 'There is a God,' and 'Killing is wrong.' Beliefs help us make judgments about information and lead to inferences about cause and effect (McNelly 1973). Beliefs offer a basis for mental pictures of what we hold to be real and acceptable (Rediker et al. 1993). So, beliefs contribute to each person's perceptions of reality. These individual perceptions do not always agree (Petty and Cacioppo 1981). An individual's internal logic truly produces their view of reality. What is unquestionable fact to one may be pure fantasy to another.

Beliefs can be conceptualized as conglomerating together to form a mental sieve through which all sensations received by an individual are filtered. Belief systems link different concepts with one another and give form and meaning to information (Rokeach 1968). They endow the holder's given state of affairs with sufficient validity and trustworthiness to warrant their reliance on their system as a guide for thought and action (Harvey 1986). So, belief systems are self-sustaining and can be expected to change slowly, if at all. Belief systems are heavily reliant on an individual's experience and are, therefore, constructivist in nature. Belief systems also help us to make predications about the future by generating expectations about situations, actions and other people (Paap 1989). In short, a belief structure acts as a template for making sense of the world and in deciding what is correct and what is not (Waern 1977; Rumelhart 1980).

Belief systems are also referred to as schema. Regardless of the term used, belief systems are not random gatherings of beliefs that remain static within an individual's mindscape. Their structure is dictated by a connectedness that is logical, at least, to the holder of the system. We are all believers in our own way, you could say. This connectedness constrain what an individual will find true about reality. Horizontal constraint is where beliefs are related to each other in a chain-like sequence. For example, beliefs on chemicals, manufacturing and big corporations tend to be linked together. Vertical constraint is where successive beliefs, often taking the form of a logical syllogism, are derived from each deeper belief (McGuire 1960; Bem 1970). Syllogisms are aspects of reasoning where one premise about something leads to a logical premise about something else and hence to a conclusion. For example, my friend Elaine reads a daily on-line newspaper to keep her informed of world events. She is always well informed. Therefore, newspapers keep you informed.

All of these assertions about beliefs might lead you to conclude that changing opinions and affecting behavior is a foolhardy proposition. Communication that directly attacks one's belief system will almost always be ignored or, worse, interpreted as threatening. And, yet consider how much of the information you come into contact with daily is meant to change your mind on one topic or another – advertising, political rhetoric, religious pronouncements and so on.

Clearly, new information often does not squarely fit with our existing belief structures. Such information may be disregarded, 'glossed over' and not mentally processed in full, or delegated to an area of uncertainty where conscious reasoning can still occur. Important in the mental processing of new information vis-à-vis belief structure is an individual's expectation about whether the incoming information needs to be fully considered or not. More on this process as we move from beliefs to their close relatives, values.

6.5.2 Value

A value is a personal benchmark of what is desirable. They depend on beliefs and are more directly responsible guiding everyday behavior. Like beliefs, values provide a framework for making decisions in specific situations (Kluckhohn et al. 1951). Whereas beliefs are truth-claims about what cannot be proven or fully attained, values offer standards by which individuals judge themselves and their realities. So, an enduring idea that a specific mode of conduct or end-state of existence is personally or socially preferable is a value (Rokeach 1973). From the belief 'Knowledge is good' could come the value 'More knowledge is better.' In function, beliefs and values are closely aligned and are nearly inseparable when you think about them in designing communication campaigns. Again, attacking values will almost always get you nowhere with recipients of your messages.

Like beliefs, values are organized into systems. A value system allows its holder to envision preferred modes of conduct and desirable end-states of existence. In essence a value system is a learned organization of principles and rules to help us choose between alternatives and resolve conflicts (Rokeach 1973). Values are standards with which we judge claims and decide whether they are worth challenging, protesting, debating or ignoring. All experiences are examined using values. Predicting personal judgments by understanding an individual's values is not exact, but can be helpful in planning environmental communication.

6.5.3 Attitude

Information about specific objects and situations is framed and assessed through one's belief and value structures. This process of framing, and the inevitable drawing of inferences about an object or situation, is how attitudes are formed. Attitudes are focused and enduring mental organizations of information, based on several beliefs and values. They predispose one to respond in a certain way (Eagly and Chaiken 1993). Notice how attitudes are a product of beliefs and values. Sometimes understanding the underlying belief and value systems can help a communicator understand what factors may be hindering members of their audience from accepting a message and making changes based on it. Because attitudes are specific toward individual things within the reality of each individual, attitudes vary subtly among members of groups.

6.5.4 Worldview

When we look at different audiences, we need to be mindful of their socio-cultural backgrounds and other aspects that define how they will look at the world. Worldviews are the composite of all our beliefs, values and attitudes that we have come to accept as our own. We all have a particular mindset that dictates how we filter information and make sense of what is happening. Our personal worldview also dictates what we expect to receive. When a message seems alien to this perspective, we will misinterpret the message, or just disregard it. This could be thought of as faulty encoding and decoding within the communication model.

6.5.5 Opinion

When one expresses a belief, value or attitude, one has an opinion. Opinions are, therefore, the verbal evidence of an individual's worldview. Whereas beliefs, values and attitudes are rarely stated openly and can be difficult to reveal, opinions are made manifest. Opinions can be thought of as beliefs, values and attitudes put into words. Without some form of communicative representation, there can be no opinion (Rokeach 1968).

6.5.6 Situational Factors

Just as noise permeates and confounds communication systems, situational factors can short circuit, derail and otherwise disrupt people's beliefs, values, attitudes and opinions. Situational factors are all of those annoyances, unfortunate events, and disasters that arise continually to complicate everyones' lives: phone calls on your way out the door, flat tires, being broke, getting the flu, a death in the family, a flood, a stock market crash, a meteor strike, and the list goes on. Even peer pressure and social norms can be situational factors. From the standpoint of a communicator, situational factors get in the way of free flow of information. They are a form of noise.

Situational factors may interfere to make a person continue in a mode of behavior that is seen as 'real-world' rather than a lofty ideal. Sometimes a message may have to empower people to make changes they would like to make, rather than just informing them of a situation.

6.6 Memes

An often heard frustration in any communications effort is when a message seems well-prepared, transmits cleanly, and is even understood clearly, yet still does not elicit an anticipated reaction. Such a message is falling on 'deaf-ears' so to speak, probably due to a conflict with a specific belief or value. This is particularly evident when one considers 'memes,' which are psychological constructs, internal representations of reality. While the term 'meme' can be traced back to Richard Dawkins (2006), Richard Brodie (1996, p. 11) profiles this concept further with a working definition: 'A meme is a unit of information in a mind whose existence influences events such that more copies of itself get created in other minds.'

Brodie likens them to viruses because they spread readily, replicate, mutate and evolve over time. Memes can be easily transferred from one individual to another through generation after generation. They are a virus of the mind. Memes are contagious ideas that take root in our worldviews, beliefs and values. A example meme might be 'economic growth is good.' There is no value judgment on that meme save the subjective value we personally place on it. A virulent meme is not one that is subjectively good or bad, merely that it spreads easily. Memes may be pathologically bad, yet spread readily and be resistant to change. The main problem with memes is that we do not usually consider them until we are shown them. It is here where critical thinking and self-reflection are crucial tools in recognizing and disinfecting against memes.

A meme will continue to exist because enough people act on it, entrenching the meme within our social identity much like a virus in established with the cells of a host. Memes are not 'bad' as such, after all, we use them for good thinking as well as pathological thinking. After all, 'recycling is good for the planet,' and 'preventing pollution promotes clean healthy air' are subjectively good memes. Yet more and more, we find people becoming more enslaved by craftly constructed sound bites from those who understand how to plant memes into other's minds. Many of our cultural memes pass without analysis. Advertisers work to promote memes by showing us ideas about ourselves that we come to accept without question. Are we really only happy when we purchase brand X or that we can only be successful if we do action Y? What really makes us happy and successful? Some people are shopaholics, but how often are they exposed to the result that consumption is wreaking havoc on the planet? Yet simply telling a shopaholic that their behavior is detrimental to environmental health is unlikely to elicit an environmentally sound action. The only way to get through is to get the person to consciously review their shopping habits and then consider alternatives. Such an interventionist message would need to be processed through a central mind pathway such that it is reflectively considered. In other words, the sender would need to be a trusted source and there would need to be ample time for message processing.

Ridding oneself of pathological memes is akin to disinfection. A solution takes critical reflection and mindful thinking to identify our own memes and the rationale behind them. In short, we encourage everyone to become individual thinkers. This does not mean to dissociate oneself from community, but rather to think for oneself with depth and intention. As Brodie (1996) says, 'Reclaim your brain.' Recognizing memes inherent in our topics and constructing our messages to promote critical reflection in our audiences is as crucial an analysis for our audiences as much as knowing who they are.

6.7 Locus of Control

Locus of control (LOC) is another psychological concept important in environmental communication. People perceive their ability to control situations in which they are involved in different ways. A person who believes their actions can affect a situation's outcome has an internal locus of control. One who believes they are powerless in changing a situation's course has an external locus of control. Internal and external loci of control can be conceptualized as falling along a continuum. While in different situations the same individual may have different locus of control reactions, one's predisposition toward one end of this continuum or the other is moderately stable over time (Rotter 1954; Levenson 1972; Phares 1976; Lefcourt 1980).

INTERNAL LOC (**I** control my situation)

- Believe their actions can make a difference.
- Internalize control.
- Account for only a small number of people.
- Information seekers.
- Take action in reference to their beliefs although such action is not legally mandated. Example: a person who practices recycling when it is not mandatory.

EXTERNAL LOC (THEY control my situation)
- Either do not have the necessary information to take action or do not believe their actions will make a difference. May just require motivating or training in action skills to become internals.
- Need to be convinced of the importance of their role.
- Believe they cannot facilitate change, only 'powerful others' can do so. Powerful others are perceived having control and does not depend on any real group being active and noticeable.

When audience analysis unveils information about an audience's locus of control, it can be extremely helpful in planning a campaign. Internals on a particular issue may simply need to made aware of a problem. Externals need much more attention if they are to become involved. A small group of externals may fall close to a cut-off point on the continuum and only need some action skill training and more information to generate activist tendencies. Many more externals will need to be motivated to act so that they gain ownership of the problems addressed. They may also need empowerment messages to become active. The group perceiving 'powerful others' exist may need extensive empowerment programs before they will even begin to consider listening or acting.

In summary, a communicator needs to understand that beliefs and values are extremely difficult to change. Indeed, most people will militantly resistant such changes. Attitudes and opinions are expressions of these factors and make observation of the underlying cultural norms easier to fathom. This overall cultural mindset determines a worldview that a communicator needs to understand if they are to place a message to get a target audience to change something they have 'always done.' When a communicator is trying to get an audience to adopt new ideas, it is critical to know the barriers that exist within that audience to adopting something that will change their lives.

Understanding how an audience 'thinks' is a crucial step in assessing how likely they are to respond to a specific message and how further messages need to be structured. As members of an audience advance up the adoption sequence, it is necessary to keep reinforcing messages so they do not reject them. Knowing why an audience is interested in the first place will help in avoiding barriers that lead to rejection. LOC is important since people are more likely to respond once they feel they have the power to do something. Apathy is a major barrier that may be overcome by establishing ownership of a situation. Work to ensure an audience feels it is empowered to act. Decoding of a message will change significantly depending on the psychological mindset of an audience. An audience that is highly empowered with an internal LOC and a adopter profile of an early majority/greenback-green may need just information and a modicum of encouragement to act. An audience comprised mainly of grousers/late majority with a LOC in 'powerful others' will need to be targeted for multiple waves of empowering and motivating messages before any hope can be expected for adopting a new idea. Helping an audience understand how new ideas will impact their lives and then reinforcing the message demands a lot of work and a long-term messaging campaign if success is expected. Knowing your audience is a major step in this process.

6.8 A Model of Citizen Participation

The model of environmentally responsible behavior put forth by Hungerford and Volk (1990) details three levels of activity a communicator should consider when developing a message for a target audience. After considering the psychological factors composing an audience's worldview, also consider each level of this model and each variable. Note how the focus of a message must also vary to address specific points of these variables. While some of these variables may be more important than others for any given audience, there is a Gestalt at play, were all the 'pieces' need be there before expectations of success are met. Like the adoption sequence, this model is another way of viewing the successive steps of communication campaigning that need to be planned if the overall goal is to move the audience to act.

6.8.1 Entry Level Variables + Ownership Variables + Empowerment Variables → Environmentally Responsible Behavior

Entry Level variables include…

- Awareness about a situation/issue
- Sensitivity towards a situation/issue
- Basic knowledge about a situation/issue
- Having developed attitudes toward a situation/issue

Ownership variables include…

- In-depth knowledge about a situation/issue
- A sense of personal investment (a stake)
- A personal commitment to resolve a situation
- Knowledge of the positive and negative consequences of decisions about a situation/issue

Empowerment variables include…

- Knowledge of action skills
- Internal locus of control
- Intention to act

Intention to act is predicated on motivation and personal needs. Communicators have some ability to affect motivation, especially with well-understood groups of adults.

Audience analysis involves much more than mere characterization. A successful communicator must also understand their audience. This might be achieved first-hand by talking to selected members of a target audience, or by gathering information on actual behaviors of a target audience. If a communicator can reveal which beliefs might be threatened by the issue being addressed, then messages can be better designed and targeted. Opinions can be examined without producing outrage. Beliefs and values have to be carefully and gently called into question or a campaign will fail due to audience resistance. In the Hungerford and Volk (1990) model of Responsible Citizen Action, the second step of 'ownership' is where reflective space may be provided for recipients to self-examine opinions and attitudes – perhaps even values and beliefs – as they bear on the issue's resolution. Once an audience member begins to feel ownership, they are more apt to move to empowerment. An owner is also more likely to work to overcome situation factors, not allowing minor annoyances to permanently distract.

Different people can conceptualize the same issue in radically different ways. Individual attitudes are narrowly focused and dependent on their immediate circumstances. Public opinion is often based on loose and shifting coalitions. Likewise, environmental attitudes tend to be issue-specific and not necessarily continuous across all other issues. Mass belief systems usually have little depth or breadth to them, and are organized around more day-to-day and apolitical concerns unless galvanized by a specific current situation. During times of controversial situations that affect large groups of people, there are clear signals given within a community on what and whose interests are affected. Environmental protection consequently has often been a coalition of 'pet peeves' rather than a unified movement (deHaven-Smith 1988). The selected attitude theories given here emphasize just how difficult it is to generalize from one population to another, or even within populations. Though it is not feasible to analyze each person in an audience, a professional communicator needs to build a campaign on an awareness of the psychological factors at play within their potential audiences. With probing and skilled audience analysis, it is possible to understand how to construct a successful message or why a message already in use may not be working. Audience analysis works hand-in-hand with evaluation. Both are crucial components to overall success of environmental communication.

6.9 Motivation

Motivated people act. Motivation can be described as an individual's drive to get things done in their life. Motivation directs formation of new opinions and transforms them into actions. As such, it is closely related to locus of control. People prefer to do what they are interested in, so messages that appeal to interest, most often appeal to the audience's motivation.

Motivation can be typed as general, specific, intrinsic or extrinsic. General motivation is an enduring disposition to strive for new knowledge and skills (Brophy 1987). When tied to a message or sender, motivation becomes specific. For example, when someone you love asks you to do something, you have more motivation than if a stranger would ask you to do the same thing. Motivation that comes from within an individual is intrinsic. Other names for intrinsic motivation are curiosity, personal gratification and growth potential.

When motivation is reinforced by external standards – test scores, salaries, degrees – it is extrinsic. Use of extrinsic motivators in communication campaigns has been shown to be risky. Meeting a standard or being compared to a larger group may motivate people to succeed, but fail to affect permanent behavioral change. Reinforcement is motivated by a fear of sanctioning and not necessarily a desire to succeed. Plus, giving extrinsic rewards for doing what someone is already interested in actually decreases their specific motivation (Calder and Staw 1975; Morgan 1984). When giving rewards, it is better to reward individual action rather than achievement of higher standards. People like to be compared with themselves, not faceless statistics derived from large groups. When using enticements as motivators, care should be taken to ensure the target audience does not become unmotivated because of misplaced efforts to promote action and learning.

The best motivation comes from within. Communication techniques that tap into people's intrinsic motivation can be highly effective. But, it is difficult to produce messages that do this (Combs and Avila 1985). Humanistic approaches state there are no unmotivated people. The sender of a message must provide the relevant connections to the receiver. Answering the question 'What's in it for me?' is key.

6.9.1 Motivational Needs Models

Our personal needs affect how much motivation we will invest in various tasks. Needs theory proponents stipulate that motivation occurs because people want to remedy deficiencies in their basic needs and to attain proficient access to higher needs or desire for personal growth. The main implication of needs theory to the communicator is that people are unlikely to listen to a message that appeals to the aesthetic or higher thinking concerns before the basic needs of everyday living have been met. For example, asking people below the poverty line to 'save the dolphins' by only buying higher-priced dolphin-free tuna is not likely to have much effect. These people most probably cannot afford to support such a boycott, and may be too preoccupied with day-to-day living to worry about anything but surviving in their own lives. Understanding your audiences 'needs' will help you develop appropriate messages. The three versions of motivational needs models briefly described here emphasize the nuances of needs that need to be understood by the communicator.

Murray's (1938)/Atkinson's (1964) Manifest Needs theory proposes multiple 'learned' needs interact simultaneously and two factors, direction and intensity, drive the desire to satisfy needs. Direction is the focus of the need; people often look for a safe haven. Intensity is the strength of desire to satisfy that need. For example, when only a mild threat exists there is little desire to hide, yet when a large threat exists there is a paramount desire to escape and hide. Some of the needs identified as manifest are achievement, competition, individuality, nurturance, order, power and empowerment. What people need is often based on cultural upbringing and cultural expectations.

Maslow's (1943) Hierarchy of Needs is the most well-known needs theory and posits that humans have a hierarchical arrangement of needs from the base (deficiency needs) to the existential (growth needs), visualized as a pyramid. These needs were expanded in 1970 to seven levels with the first four deficiency needs being Survival (shelter, food, water, and warmth), Safety (freedom from physical or psychological threat), Belonging (love and acceptance from others) and Self-esteem (recognition and approval, self-worth). The top three growth needs are Intellectual Achievement (knowing and understanding), Aesthetics (order, truth, and beauty) and Self-actualization (philosophical thinking, spirituality). The main upshot of Maslow's theory is an individual must first satisfy more basic needs before they will advance to the higher levels of need.

Alderfer's (1972) Existence/Relatedness/Growth (ERG) Needs theory includes two extra components of 'frustration-regression' and 'satisfaction-progression' which control levels of need at which a person remains. Therefore, a person acts on needs that are under frustration since these needs take precedence over ones that are already satisfied. When a need is satisfied, the person will want to progress further. If a need is frustrated, then a person will regress and prioritize that need until it is satisfied again. This does not imply that other needs are disregarded, rather it suggests frustration needs will be answered before other needs.

All three motivational needs theories emphasize individuals' requirements that their needs be met before they will entertain change. In segmentation of groups, early change agents, such as innovators, early adopters or true blue-greens, could be equated to those operating at higher and already satisfied needs levels. This allows them to contemplate becoming involved in new situations. They are not preoccupied with their basic needs. Alternatively, those categorized as laggards or basic browns may be more concerned with everyday needs and are unlikely to find time or energy to devote to esoteric issues, which they also feel unempowered to affect anyway. If a specific unempowered audience is important to the outcome of a communication campaign, then the communicator will need to address that audience's needs before attempting to get them to begin adopting new behaviors.

6.9.2 How to Motivate Adults

With all of the motivational theory to work from, here are encapsulated directions for reaching adult audiences through your environmental communication campaign. (Reaching children requires different considerations, both practical and ethical.)

- Create or identify a current need in the audience and then respond to it.
- Get them to develop a sense of responsibility.
- Create and maintain interest within the audience.
- Structure messages to apply to real life.

- Give recognition, encouragement and approval if they demonstrate action or interest. Praise and constructive criticism are best, rather than just praise.
- Foster wholesome competition and cooperation.
- Get excited yourself and then share your enthusiasm.
- Explain what is in it for them.
- See the value of internal motives.
- Intensify interpersonal relationships.
- Give them a choice. Let them be part of the process.

6.10 Consumerism as a Way to Understand Preferences

In *Sustainable Capitalism*, Ikert (2005) gives a potential Bill of Rights for Sustainability. One of the rights he proposes impacts environmental communication heavily: 'The Right of Individual Thought and Expression – The right to accurate and unbiased information to prevent subversive influence of one's thoughts for economic purposes.'

It is not our purpose here to comment on how advertizing and consumerism works, but rather that a basic understanding of the business forces that drive consumerism should be understood as a focus for understanding audiences. It is believed that consumers have free choice (real or imagined) in the decisions they make about products they buy, or more importantly, the image or lifestyle portrayed by the advertizing that works to promote consumer decision-making. This obviously is affected by the quality and accuracy of information they receive. Just as in advertizing there is a bias to promote one perspective (the product being advertized), in other communications there is often a bias to promote one point of view for persuasive purposes. Just as people make choices on products they buy, so they make similar choices on messages they will attend to, or ignore. In response to advertizing, people make choices based on how a message 'speaks to them' – if it is not focused on what they 'want' to see they are unlikely to read or listen. A lot of effort needs to be placed on analyzing one's audiences, and then to develop a thorough message that is accurate and minimally biased. It might be said, our goal as communicators is to help our audiences make good rational and critically informed decisions – not tell them what to think.

What happens when consumers are allowed to be fully informed about the process driving consumerism? Is there evidence this is helpful in making rational decisions on whether to purchase or not? A buying decision might concern something as straightforward as a garment made in a third-world 'sweat-shop' or an item made from wood harvested unsustainably in a tropical rainforest. Kysar (2004) sadly finds only a tenuous connection between consumer choice and the process that goes into a products manufacture. Simply put, consumers overall, do not seem to care much. Much more research is needed into this potentially crippling aspect of consumer psychology and message effects.

6.10.1 *Business Communication to Assist Consumer Choice*

Polls show consumers desire applicable information to make informed decisions about products they purchase. This readily suggests that there exists some form of shared responsibility for the communications process. Some multinational corporations have now begun to be more open and ready to share to keep consumers informed. It is this informational exchange that helps consumers understand how their actions affect the environment, and through choices in buying habits have sent feedback that influence business behaviors. An example of one way that consumer-based preferences have been manifested in recent years is through corporate sustainability reports (see Section 20.5).

The Dow Chemical Company for example does not sell products directly to consumers, yet is a global supplier of thousands of products to other sectors of the business community. Dow now recognizes that they need to engage the full value chain from their point of control to the consumers. Dow is now collaborating with other companies to develop relevant information for institutions and consumers to consider in their purchasing decisions.

When considering a user's perspective several factors must be considered not least whether a product does what it was meant to do if it does it safely. The product also must be cost-effective from the user's perspective. Added to supply and demand, there are other considerations such as environmental impacts (water, emissions, hazards, energy, green house gases, etc.) and social considerations (worker safety, basic human needs for food, water, shelter and health, etc.). For sustainability, these considerations need to be assessed over the product's full life cycle – this is part of a new way of thinking called 'Industrial Ecology.'

Dow's corporate communications strategy has been to make their chemical labels look more like food nutrition labels because on a food nutrition label, consumers are provided information on a variety of factors (sugar, protein, fat), various desirable nutrients (vitamins, minerals) and some to be avoided (sodium, saturated fat). A relatively new but blossoming field is 'green chemistry.' With the new chemical labels each person can make a better choice on the use of a chemical used in a process. Dow has stated, 'It is critical to define the components and metrics for such a label, and the process should engage all of the stakeholders in its development' (Consumer Choice 2009).

Ideas presented in this chapter emphasize the importance of honestly knowing your audience. Successful communication campaigns are built on such familiarity. When you know what level of knowledge and social action the receivers of your messages have, you are more likely to succeed. You should be able to explain their beliefs, values and attitudes, and the opinions they express. With such understanding you can motivate your audience. Environmentally responsible behavior can result. Know them well and you will get the response you expect.

6.11 Case Study: Audience Analysis. Environmental Radio Soap Opera for Rural Vietnam

With financial support from the World Bank, an environmental radio soap opera – 'Que Minh Xanh Mai' ('Forever Green My Homeland') – was produced and broadcast to Vietnamese rice farmers in Mekong Delta, the country's 'rice basket.' Called 'education-entertainment' by its producers, the radio drama aimed to inform farmers on wildlife conservation principles, reductions in pesticide use, and straw burning and water use, so as to protect ecosystem services. The 104 episodes of the soap opera are based on rural settings and feature daily struggles, joys and loves in rural life. The show was written by a team of technical experts ('the turtles') and creative writers ('the peacocks'). The series was broadcast twice a week. On-the-ground listeners' clubs were also organized, which had 20–30 members each and met monthly to discuss entertainment and educational content of previous episodes. A meet-the-audience day was held, where attendees were asked to guess who played the different characters. The messages and media for this highly successful campaign flowed from an audience analysis conducted in 2005. It showed TV and radio remaining the main sources of technical information in rural Vietnam. The extensive analysis began with a stakeholder workshop, followed by focus groups, and then resulting data were compiled alongside a literature review. All despite a similar radio drama series airing the 2 years prior, which reached two million Vietnamese and precipitated demonstrable shifts in integrated pest management practices. Even though TV was more popular with the target audience, the analysis pointed toward radio as the medium of choice, because farmers were more likely to be attending to chores and sharing the listening experience with others, making the receipt of the environmental messages a social activity and pertinent to the work being attended to in real-time. The analysis also illuminated knowledge gaps for incorporating into the show's scripts. For instance, the production team learned that 90% of the farmers burned their left-over rice straw, while a conservation goal is to have farmers reincorporated the straw back into the soil.

Credit: Environmental Radio Soap Opera for Rural Vietnam.

References and Further Reading

Alderfer CP (1972) Existence, relatedness and growth. The Free Press, New York
Atkinson JW (1964) An introduction to motivation. Van Nostrand, New York
Baron RA, Branscombe NR, Byrne DR (2008) Social psychology, 12th edn. Allyn & Bacon, Boston, MA

Bem DJ (1970) Beliefs, attitudes and human affairs. Brooks/Cole, Belmont, CA
Bermejo F (2007) The internet audience: constitution & measurement. Peter Lang, New York
Bird SE (2003) The audience in everyday life: living in a media world. Routledge, New York
Brodie R (1996) Virus of the mind: the new science of the meme. Hay House, New York
Brophy J (1987) Syntheses of research on strategies for motivating students to learn. Educ Leadership 45(2):40–48
Brown S, Christensen CM, Deighton J, Dolan RJ, Fader PS, Fournier S, Gourville JT, Moe WW, Sarvary M (2002) Understanding consumer behavior. Business fundamentals series
Calder B, Staw B (1975) Self-perception of intrinsic and extrinsic motivation. J Personality Soc Psych 31:599–605
Combs A Avila D (1985) Helping relationships, 3rd edn. Allyn & Bacon, Boston
Consumer Choice (2009) Green Chemistry Initiative, California Department of Toxic Substances Control, Sacramento. http://www.dtsc.ca.gov/PollutionPrevention/GreenChemistryInitiative/upload/Dow_Consumer_Choice.pdf
Dawkins R (2006) The selfish gene: 30th Anniversary edition. Oxford University Press, New York
DeHaven-Smith L (1988) Environmental belief systems: Public opinion on land use regulation in Florida. Environ Beh 20(2):176–199
Eagly AH Chaiken S (1993) The Psychology of attitudes. Harcourt, Brace, & Jovanovich, New York
Escalada M et al (2006) Audience analysis report, environmental radio soap opera for rural Vietnam. World Bank and International Rice Research Institute, Manila, The Philippines
Harvey OJ (1986) Belief systems and attitudes toward the death penalty and other punishments. J Personality 54(4):659–675
Hungerford HR, Volk TL (1990) Changing learner behavior through environmental education. J Environ Educ 21(3):8–22
Ikert J (2005) Sustainable capitalism: A matter of common sense. Kumarin, Bloomfield, CT
Kluckhohn C et al (1951) Values and value orientations in the theory of action. An exploration in definition and classification. In: Parsons T, Shils EA (eds) Towards a general theory of action. Harvard University Press, Cambridge, MA, pp 388–433
Kysar DA (2004) Preferences for processes: the process/product distinction and the regulation of consumer choice. Harv Law Rev 118(2):525–642
Lefcourt HM (1980) Locus of control and coping with life's events. In: Staub E (ed) Personality: basic aspects and current research. Prentice-Hall, Englewood Cliffs, NJ, p 229
Lefcourt HM (1982) Locus of control – Current trends in theory and research, 2nd edn. Lawrence Erlbaum, New Jersey, p 186
Levenson HM (1972) Locus of control and other cognitive correlates of involvement of anti-pollution activities, Dissertation. Claremont Graduate School, California
Lionberger HF (1960) Adoption of new ideas and practices. Iowa State University, Iowa
Maslow AH (1943) A theory of human motivation. Psych Rev 50:374–396
Maslow AH (1970) Motivation and personality. Harper and Row, New York
McGuire WJ (1960) A syllogistic analysis of cognitive relationships. In: Hovland CI, Rosenberg MJ (eds) Attitude organization and change. Yale University Press, New Haven
McNelly JT (1973) Mass media and information redistribution. J Environ Educ 5(1):31–35
McQuail D (1997) Audience analysis. Sage, Thousand Oaks, CA
Morgan M (1984) Reward-induced decrements and increments in intrinsic motivation. Rev Educ Res 54:5–30
Murray HA (1938) Explanation in personality. Oxford University Press, New York
Paap KR (1989) Applied cognitive psychology. In: Gregory WL, Burroughs WJ (eds) Introduction to applied psychology. Scott Foresman and Co, Glenview, IL
Phares EJ (1976) Locus of control in personality. General Learning Press, Morristown, NJ
Petty RE, Cacioppo JT (1981) Attitudes and persuasion: classic and contemporary approaches. William C. Brown, Dubuque, IA
Perloff RM (2007) The Dynamics of persuasion: communication and attitudes in the 21st century, 3rd edn. Lawrence Erlbaum, Mahwah, NJ

References and Further Reading

Rediker KJ, Mitchell TR, Beard DW, Beach LR (1993) The effects of strong belief structures on information-processing evaluations and choice. J Beh Decision Making 6(2):113–132

Rogers EM, Shoemaker FF (1971) Communication of innovations. Free Press, New York

Rogers EM (1995) Diffusion of innovators, 4th edn. Free Press, New York

Rokeach M (1968) Beliefs, attitudes and values. Jossey-Bass, San Francisco

Rokeach M (1973) The nature of human values. Free Press, New York

Roper (1990) The environment: public attitudes and individual behavior. The Roper Organization, New York

Rotter JB (1954) Social learning and clinical psychology. Prentice-Hall, Englewood Cliffs, NJ

Rumelhart DE (1980) Schemata: the building blocks of cognition. In: Spiro RJ, Bruce BC, Brewer WF (eds) Theoretical issues in reading comprehension. Lawrence Erlbaum, Hillsdale, NJ, pp 33–35

Solo RA, Rogers EM (eds) (1972) Inducing technological change for economic growth and development. Michigan State University Press, Michigan

Tewksbury D (2005) The seeds of audience fragmentation: specialization in the use of online news sites. J Broadcast Electron Med 49(3):332–348

Van Rees K, van Eijck K (2001) The fragmentation of the media audience. In: Schram DH, Steen G (eds) The psychology and sociology of literature. Benjamins, Amsterdam

Waern Y (1977) Comprehension and belief structure. Scand J Psych 18:266–274

Webster JG (2005) Beneath the veneer of fragmentation: television audience polarization in a multichannel world. J Commun 55(2):366–382

Yopp JJ, McAdams KC, Thornburg RM (2009) Reaching audiences: a guide to media writing, 5th edn. Allyn & Bacon, Boston, MA

Youga JM (1989) Elements of audience analysis. MacMillan, New York

Chapter 7
Evaluating Your Messages' Effects

7.1 Introduction

Evaluation is the on-going process of providing feedback about the effects of your messages. Just like audience analysis and issue investigation, evaluation is indispensable to carrying out your communication plan well. Yet, evaluation is often the most neglected portion of communication campaigns. Many programs and plans often have no basis for determining how successful they were or for analyzing where problems may have occurred. Imagine attending a class where the only measure of success is when the teacher feels satisfied that they have taught well. Does that give any measure of how much students have learned, relate good techniques to other teachers, or help the teacher improve instruction?

Many communication campaigns have been set up with just this sort of vagueness in order to justify themselves. Does knowing how many fact sheets were taken at a local nature preserve during a given weekend tell us how effective the fact sheets actually were in imparting ecological information to visitors? Do users actually read the fact sheets? How do we know? Answers to such lines of questioning are the essence of evaluation. An environmental communicator needs to genuinely want information that helps them understand problems and improve programs.

Think back to a communications model again. Evaluation provides formal feedback in an environmental communications system. Evaluative feedback is dynamic and can take place at most any stage as the system operates. In this chapter, we examine the need for feedback in the form of evaluation – why it is needed and how to conduct it.

7.2 Purposes of Evaluation

Evaluation is a systematic practice of judging program effects. It measures worth and effectiveness of what is being done. A program in environmental communication could be an interpretative talk, a classroom activity, or a mass media campaign.

Environmental communication messages seek to modify an audience's environmental behavior in some way. Whatever the particular modes employed by your program, there is embedded within it a need to check if it is working the way you supposed it would. A program that is not meeting expectations can be modified or terminated. Evaluation has at least three other purposes worth considering. They are:

- To assist decision-makers who are responsible for deciding policy. People with policy-making duties need to know how effective a program is in order to wisely make decisions about future resource allocations. Such decisions can be about the continuation of the program itself or about broader policies.
- To create accountability for resources used. When resources have been committed to a program, the communicator, as the program manager, needs to track and justify how those resources were used. Resources include budgets, personnel, equipment, and, also, energy and time. Resources, especially cash, are always limited. Decisions about their outlay need to be based on valid information generated from defensible evaluation.
- To serve a political function. This is probably the most surreptitious use of evaluation. A simple equation is: Evaluation = Information = Power = Politics. Politics is defined here as the art of exercising power in competition for resources and setting of priorities. The idea of politics covers both public and private domains of human society. There are elected officials and corporate agents with tremendous power. Simply stated, evaluation of communication campaigns can give a political player an upper hand in some situation of consequence. High-quality evaluation is timely and pertinent to this exercise. The people who provide such information have power to help or hinder their politicos. Providing the right information at the right time can make or break a program.

7.3 Methods of Evaluating

Data for program evaluation are collected through several methods. The method selected will depend on specific informational needs behind the evaluation. Cost also comes into play in deciding the way data will be gathered. When evaluative data are arranged and put into context, they become valuable information. Communication without evaluation is haphazard and, ultimately, wasteful of resources.

Here are common methods of evaluation:

7.3.1 Surveys

Surveys use questionnaires or interviews to gather data. Whether they involve pencil and paper, computer, telephone or face-to-face dialogue, surveys are one of the most complex, misused and misunderstood methodologies of systematic information gathering. Yet surveys can yield highly significant data.

A problematic area is deciding who to survey. Selection of the group to be surveyed, called sampling, raises many statistical questions. Randomness must be present in the selection process for your results to be generalizable to any larger audience. There are many ways to sample correctly, but there are even more ways to select a faulty sample and reach false conclusions. Still, no sample is perfect. An entire population would have to be polled to absolutely know.

Developing a reliable and valid survey takes much more than just jotting down a list of questions that seem relevant and throwing them out to a group of audience members. Survey research requires careful structuring, testing and administration. The aim is to ask questions that are 'valid' (immune to misinterpretation) and 'reliable' (able to produce consistent responses). When pondering use of a survey, consider the variety of types available.

- **On-Line Surveys** – Internet-based surveys have roared into widespread use. Through services such as SurveyMonkey, Zoomerang, and Survey Methods, Web-based surveys have become easy to write and deploy. These services provide attractive formats for your questionnaires, collect data for you, and offer a battery of analysis tools at your fingertips. Speed and ease can be double-edged swords though. Valid and reliable surveys of adequate sample populations rarely result from a slap-dash instrument sent out to every email address one possesses. Drafting and revising your questionnaire is a plodding process. Use caution in instrument creation on-line; just because you can send out your survey with the press of a button, does not mean you should.
- **Mail Surveys** – A written questionnaire is sent via mail to a sample from a target audience. The biggest problem is to motivate those in the sample to fill-out and return your questionnaire. Those that do respond cannot be assumed to be representative of the whole audience, without further checking into those who did not respond and why they were non-respondent. While techniques exist for dealing with problems of non-response, it is essential to recognize the limitations it imposes on a mail survey.

    Another consideration of mail surveys is their self-administered nature. Those in your sample are asked to review their exposure to your messages and respond to questions about their reactions. This is a difficult feat to accomplish. People need to see the utility in helping you. Most times, you'll wish to provide them an incentive, some reward for filling out your questionnaire. Even with incentives, it may be difficult to get more than 50% of the individuals to respond. There is also a problem of determining if there is a difference between those people who respond to the questionnaire and those who do not. For example, it may be that those people who do not like or agree with the message, will choose not to respond.
- **Telephone Surveys** – Your sample is called using the telephone and, if they agree to participate, questions are asked directly to them. This method suffers most of the same drawbacks as mail surveys. Plus, utility of phone surveys has declined precipitously as more consumers switch to wireless phones and telemarketing became such an annoyance as to prompt 'Do Not Call' laws.

Telephone surveys can also be expensive, relative to other methods. It may remain somewhat effective with older populations who know the surveying entity well.
- **Face-to-Face Surveys** – Members of your audience are intercepted and questioned at a selected location. For example, this may occur at a museum gift shop, at the exit of a visitor center, or on the street in a mass media market where you have been running television advertising. The evaluator selects members of the audience to interview and then asks them to answer some questions. As with telephone surveys, a polite, pleasant and professional tone in which you respect the time and patience of the respondent will go a long way in motivating them to respond to you. This method can gather a lot of information in a short amount of time.
- **Personal Interviews** – Similar to on-the-street surveying, except this technique uses fewer people in a sample and asks them more in-depth qualitative questions. Individual in-depth interviews often last between 30 min and 1 h. Selected interviewees should still be representative of the target audience, though randomness will be more difficult to achieve. In-depth interviews are costly and time consuming, yet can yield superb insight on reactions to messages, issues and policies.
- **On-Site Testing** – Small groups from the target audience (usually no more than 25 people) are brought together to review various messages being developed within a plan. The messages need be set up in situations similar to final delivery. In trying to decide on the effectiveness of different kinds of messages, it may be necessary to avoid telling the audience about each message they will be viewing. This will avoid some members of the audience giving responses about what they perceive you want to hear. For example, a public service announcement may be embedded in a half-hour television or radio program. In order to get useful information about the effectiveness of the announcement, it would be useful to not tell the test group about this message, and instead concentrate on the program in which it is embedded, and see if they picked up on your embedded message.
- **Comment Forms** – Simple written questionnaires, usually formatted onto a post card or on-line response box, can be left at strategic locations to be noticed by interested members of your audience. Many restaurants will leave a comment card at each table to garner simple feedback from customers about quality of the food and staff. Similarly, comment forms left near the exit of a visitor center can give quick information on how visitors liked or disliked the center's exhibits and activities. Guest registers serve the same function at many sites.

7.3.2 Participant Observation

Sometimes, much can be learned by watching what a person does, rather than having them tell you about themselves. The evaluator takes on the role of a visitor, viewer, reader or other participant. This allows data to be gathered from the perspective of the audience. The evaluator discretely observes audience behavior and also notes their personal reactions. This is a particularly useful technique when one seeks to compare beliefs, values and attitudes, with observable behavior.

An example of this might be in a visitor center where the observer records how people look at the exhibits, how long they stay at each station, and other relevant information to gauge just how effective the exhibits seemed to be in holding the audience attention. Later interviews could be employed to give results on why some exhibits held attention while others were just dismissed. Another scenario might be where a community has begun a curbside recycling campaign. The observer might drive through the target neighborhoods on trash collection day and gauge how much trash versus recyclables was left by residents. Later interviews could yield insights into why some people recycled a lot and others very little.

7.3.3 Interviews

Studies of the adoption sequence have shown the existence of innovators who consistently try new ideas before most others. They are the vanguard of the innovation ranks. Finding and asking members of this population segment about their reactions to your messages can be a wise move, if you are attempting to shift public opinion. Here are three ways to consider discussions with such opinion leaders:

- **Opinion Leader Interviews** – Many government officials and business leaders can provide relevant information about your issue, target audiences and messages. Often, they can be interviewed to elucidate needed information. If many key designates are desired, then one of the following consensus methods may be appropriate.
- **Gatekeeper Review** – Gatekeepers are the key people in organizations who decide what information is released to the organization's members or constituents. Prime examples are the editor of a newspaper or the communication/human resources director in a large company. Gatekeeper review involves soliciting gatekeepers for their perceptions of your messages. If gatekeepers do not respond well to a message, the target audience will never get a chance to receive it. For this reason, it is wise to include gatekeeper review as a method of evaluation.
- **Discussion with Key Informants** – Key informants are people in a community or organization who have unique insight into your audience or situation. Such people can be interviewed to gain specific insights into a program or issue. Examples of the sorts of people in this group are the owner of a local store, a street vendor, a local restaurant owner, a local barber or hairdresser, and regulars at a local coffee shop or bar. Key informants are not always the 'leaders,' yet in their own right are respected by the rest of the group for their wisdom and understanding.

7.3.4 Group Consensus

When a small group that reflects your intended audience is brought together, either actually or virtually, and is guided toward consensus by a skillful facilitator, group consensus technique is being practiced. Group consensus technique is useful in needs analysis or problem definition situations. Here the group can provide insights into what needs to be evaluated or to help define the problem more succinctly. Subsequent evaluation can then be more focused on resolutions to the problem. Other uses of group consensus are for on-going reviews of a communication project so that modifications can be made, or simply to do a final evaluation of a project using a representative group.

Here are four group consensus techniques to add to the environmental communication toolkit:

- **Brainstorming/Focus Groups** – Brainstorming technique works on the Gestalt principle that the sum of ideas from individuals interacting in a group is greater than the sum of the ideas generated by individuals alone. Brainstorming works best with around 10 individuals who are representative of the target audience. While there is research suggesting brainstorming is not the most effective method for generating ideas, such a session can be a fruitful, low-cost way of generating ideas about message effectiveness (Diehl and Stroebe 1991). In most cases, if a group has experience interacting with each other already, then brainstorming tends to be an enjoyable experience that stimulates a great variety of ideas.

To facilitate a brainstorming session, follow these steps:

1. Have all participants express their ideas, as you write them down for all to see. It does not matter how silly or crazy an ideas is, for the more ideas that are expressed, the better the process works. Save evaluation of particular ideas for later.
2. Review the resulting list stressing that all ideas belong to the group. Encourage additional ideas stimulated by this review.
3. Discuss each idea, without reference to its originator. By focusing on relevancy to the issue at hand, the list can be prioritized. Take the best options and integrate them into your communication planning.

When 10–30 individuals are convened as a group and guided through a discussion by a facilitator, you have a focus group. The facilitator allows the group to talk freely about the message, channel and format. Focus group interviews have potential to provide insights into the target audience's perception of the issue and message effectiveness. Care must be taken, however, not to interpret findings of a focus group interview in a quantitative way. For example, if five out of ten of the interviewees feel a policy is unacceptable, it does not mean that 50% of the target audience will agree. The process used to deal with a focus group can be brainstorming if group members are comfortable with each other. Otherwise, the following techniques may have to be used.

- **Nominal Group Process** – This is possibly the most efficient small group method for generating useful ideas. Individuals work alone initially and then

come together as a group. This allows people to develop their own understanding of the issue at their own pace without distractions. Then, as a group, they are facilitated through these steps:
- Individuals write responses.
- Each person states one idea, which is recorded by the facilitator for all to see. Do not allow judgmental statements during this procedure.
- Participants are encouraged to ask clarifying questions and to suggest combinations of similar items.
- A key word in each remaining item is underlined. Each item should have a different key word.
- Each participant selects the most important item from the list and writes that item's key word on an index card.
- Each participant ranks the resulting key words in order of importance to themselves.
- Rank orders are averaged so a democratically generated list results. This list should be checked by the group, with the help of the facilitator, to assure that it reflects group consensus.

In nominal group process, no one openly discusses the value of any ideas. In this way, antagonistic interactions are removed from the discussion even as ideas are being judged and ranked.

- **Delphi Technique** – In some cases, it is not feasible to convene a group in a single location. If the people from whom you want to gather information are distant from each other or cannot find time to meet, then the Delphi technique offers an excellent alternative. In other cases, you may wish to not have people come together because of polarized opinions or a need to remain anonymous. Again, Delphi is a good choice. This technique avoids the problems of intimidation, co-optation, and peer pressure.

In the Delphi method of group consensus building, the evaluator drafts a series of questions about a topic, issue or program. This list is forwarded to each member of the group for their review, input and return. The evaluator then summarizes these responses and returns them back to the Delphi group for further feedback. This send-review-return-compile cycle continues until the evaluator is satisfied that some consensus has been reached.

7.3.5 *Secondary Analysis/Case Study*

Many times you can search for and find another program addressing similar issues to your own. Though it will be in another location or confronting a different situation, the details and travails encountered during planning and implementation are likely to correlate with your own. By reviewing what techniques were used and what findings were derived you may save a lot of time and money. Such a technique is called secondary analysis or case study.

7.3.6 Professional Judgment/Expert Opinion

Experts, via opinions and professional judgments, are widely used in legal, business and scientific matters. Their use is increasing in communications. Whereas opinion leaders and key informants come from within a target audience, outside experts may be employed to apply their professional judgments on the merits of your campaign. Review by experts calls for rendering of professional judgments, based on experience, knowledge and some sense of the target audience. Experts may provide valuable advice to a project on ways to adapt to specific settings and circumstances. They may also prove a waste of resources. Selection of experts may introduce its own form of error or bias into an evaluation, as project managers base a decision on their own opinion of the quality and fit of an outsider to the campaign. Experts consult on their area of expertise, and care must taken to keep their consultation within that area. When using any 'hired guns,' attend to keeping their opinions on target.

7.4 Quantitative Versus Qualitative Techniques

One's training may emphasize the preferential use of either quantitative or qualitative techniques of analysis. Without going too deeply into the pros and cons of each, it is sufficient to say that any technique used should be based on the questions needing to be answered. What are you trying to find out? One paradigm is not better than the other, but depending on the data needed for an informed evaluative finding, the type of technique employed needs to be carefully considered (Campbell and Stanley 1963; Creswell 2008). Einstein is credited with saying, 'Not everything that can be counted counts, and not everything that counts can be counted.'

If the purpose of research, including evaluation, is to become more familiar with phenomena, such as gaining new insights about a program or how an audience is responding, then the end product can said to be exploratory. When exploring, we are not looking for numbers so much as understanding. If we want to portray accurately the incidence, distribution and characteristics of a group or situation; then the end product is description. When describing, we want to investigate the relationship or association between characteristics and the purpose is to explain and/or predict the things under evaluation. In this latter case we may even employ experimental designs (or even quasi- and pseudo-experimental designs) (Campbell and Stanley 1963), using hypothesis testing. If your evaluative needs are complex, consider linking up with an expert. Commonly, communicators consult with their local college or university or enlist expertise for their extended professional network.

If you do decide to go it alone, there are numerous evaluation designs from which to choose. Hansen (2005) summarized many of the models and places them into six broad categories: results models, process models, system models, economic models, actor models, and program, and makes recommendations on which are better to use based on specific needs of the evaluation – its purpose, the object or problem to be solved. We recommend the Bennett (1976) model of evaluation as a framework.

7.5 Types of Evaluation

Evaluation takes place throughout the course of an environmental communication campaign. Timing of your assessments can be used to classify four types of evaluation based on when the results are taken and applied. This distinction is taken from the Bennett model of evaluation (Bennett 1976). Although this evaluation model is widely used by many in the environmental communication field, it is only one of many available models (see Hansen 2005). Here are the Bennett process types:

7.5.1 Formative

Evaluation that takes place during a campaign's design and early in its implementation is labeled formative. Such evaluation is meant to gain information about the audience, head-off potential problems and help confirm correct channels have been selected. There are several subtypes of formative evaluation, including:

- *Context* evaluation, which assesses the information preferences and needs of the target audience.
- *Input* evaluation, which assesses which channels and formats are most appropriate to reach a target audience.
- *Program* evaluation, which addresses the effectiveness of actual transmission of the message before it is sent out to the main audience. This is also referred to as *pretesting*.
- *Needs assessment* is where needs are identified and prioritized for further analysis.
- *Needs analysis* is where identified needs are analyzed and solution strategies developed.

7.5.2 Process

While a program is in full swing, process evaluation can be used to learn about effects on the fly. That is not to say resulting information is light and fluffy, however. Process evaluation allows rich and full descriptions to emerge as all aspects of a program are scrutinized. One benefit of process evaluation is problems uncovered have the chance of being remedied before a campaign is completed. The major drawback of process evaluation is its cost. Continual assessment of a program can become the most expensive part of your budget.

7.5.3 Outcome

As soon as a program wraps up, outcome evaluation can be conducted. It gives quick answers to questions about program effectiveness. How did it go? Did it make a difference? Outcome evaluation produces everything from simple head counts of the people reached by your messages to in-depth studies of shifts in audience beliefs and behaviors. Usually this more sophisticated information is required to tell if your objectives and goals were met. An excellent outcome evaluation usually compares posttest data to a pretest done during formative evaluation. This allows comparisons to be made and shows what changes have occurred within the audience.

7.5.4 Impact

Long-term effects of an environmental communication campaign are revealed by impact evaluation. This type of evaluation needs to take place after completion of the program, often after several weeks. It measures the durability of the effects. Whereas outcome evaluation might show a large shift in audience perception, impact evaluation will determine if the change was sustained. If support is shown to have fallen substantially, a new program of simple reinforcement of the original message might be in order.

7.6 Factors Influencing Evaluations

One of the first questions to ask as you begin evaluating your program is 'What kind of data is needed?' Do you need 'quick and dirty' results to help make a simple decision? Or will a 'thick, rich description' of what is happening be all that can adequately address your concerns? Coupled with these informational needs are the cost of the evaluation and the required timing of the results. These factors will influence the form and shape of your programmatic evaluation.

7.6.1 Cost

You cannot do what you cannot afford. Evaluation can become expensive. If the information you wish you had is financially out of reach, what cost-effective alternatives will provide similar information?

7.6.2 Expertise

Evaluation is a discipline unto itself. There are professional evaluators that specialize in complex projects of assessment. Do you have the skills to properly conduct the evaluation necessary for your campaign? If not, where do you get this know-how? You might hire an evaluative specialist or enlist faculty at your local university or college who might have an interest in the situation you are trying to evaluation. This is also a great form of civic engagement (Soska and Johnson Butterfield 2005; Watson 2007). Alternatively, it might behoove you to have one or more of your staff undergo some evaluation training. Many organizations offer regularly scheduled workshops.

7.6.3 Risk of Failure

Before enacting any evaluation within your program, be sure that all those who will use your findings really want to know how effective the program is. Do not ask questions about awareness, influence and behavior change if you do not want to know real answers. Elaborate, well-crafted and expensive campaigns sometimes fail miserably. If finding out that such a fate has befallen your project is more risky than not having answers to questions of effectiveness, do not bother evaluating. Recall one of the purposes of evaluation is 'to serve a political function.' Know beforehand how poor performance would jeopardize your standing with the powers to be.

7.6.4 Sample Make-Up/Selection

Talking to the right people is important in most any endeavor. Evaluations are no different. There are statistical concerns as well as social ones in selecting your sample. Try to avoid upsetting members of your audience by leaving them 'out of the loop.' Involving many people throughout all aspects of the program is a good way to make them accept ownership of the program. Appreciate, too, that selection takes resources, especially time, to accomplish. Finding the right people takes effort.

7.6.5 Utility

Results from evaluation efforts must be used or they are just waste of time, effort and money. While an elaborate evaluation may be desirable and the funds to perform it are available, it is critical to know from the outset just how the results will be used.

7.6.6 Timeliness

Evaluations should be planned with deadlines attached. Getting information in time to act on it, if necessary, is the real value behind evaluation.

7.6.7 *Autonomy*

Because evaluation can just as readily show resources are being wasted on a program that is accomplishing nothing as it can success stories, evaluators must be careful to have autonomy. Independence from influence is paramount to any research effort, and evaluation is a form of research. As an evaluator you have to be equally prepared to share positive and negative results. You generally will have little problem when your results affirm a program is functioning as planned. But, in evaluation, especially conducted by those within an organization, you may feel pressure to find the 'right' results. Be cautious about such intimidation. Keep your integrity and report your findings honestly, regardless of the influences on you to do otherwise.

Evaluation can tell you many, many things about your program: what works, what does not, why your campaign succeeded, why it failed, who heard what you had to say, who did not, etc. This feedback can make you a better communicator, if you pay attention to the lessons your findings offer. Evaluation, though often neglected by communicators, makes your program more valuable by heightening its accountability and giving you solid evidence of what it has accomplished.

7.7 Evaluation Plan

In our schematic of the evaluation process, the top left corner shows stakeholders and purpose. Stakeholders are the people with a vested interest in the evaluation (Fig. 7.1). It is here that primary decisions are made about the evaluative process. This drives the kind of information required to satisfy the need. Sometimes a needs assessment and needs analysis are the first steps to identify what is required to be evaluated. The next related step is to decide the purpose of the evaluation. This involves deciding who wants or needs the information and why is it needed. The project's political situation needs to be considered here. Once the process is decided, it is necessary to ponder on the mechanics. What kind of evaluation is needed? Does it require formative and/or summative evaluation? What methods will you use to collect your data? Limited resources may disqualify some methods as too expensive. Examples of this are the budget available, and the expertise of the personnel in being able to collect and analyze the information. The decision on what to evaluate then necessitates setting criteria for what information you need to collect. Setting objectives for the evaluation will guide the process. Data collection can be difficult and they can be even harder to analyze. If the type of data you require needs complex statistical analysis, will an outside expert be required? This brings us back to another review of resources. Once data are collected, the drafting of the report is critical. Is it going to be finished in time for stakeholders and decision-makers to use? If the evaluation is too complex it may take time to analyze and write a report. If an early decision was needed, then the report is redundant and the evaluation a wasted effort.

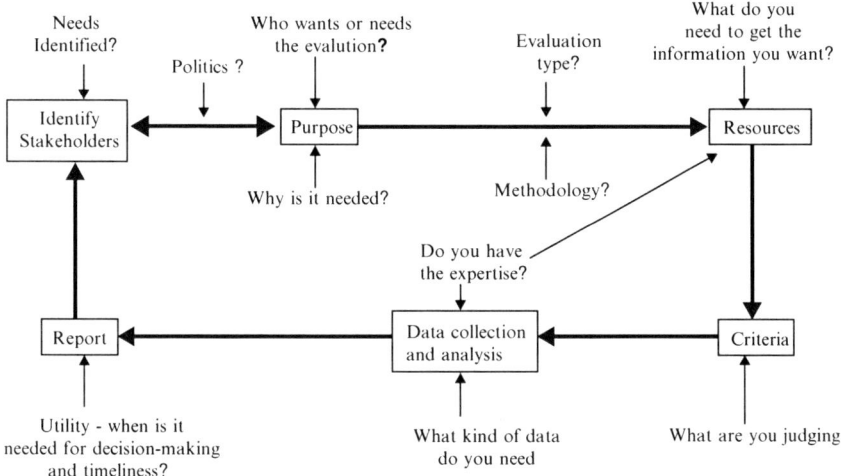

Fig. 7.1 Evaluation planning schematic. Original graphic, Richard Jurin

By moving through the different aspects of this schematic the communicator will be able to identify key components that could hinder the evaluation process. It is always pertinent to remember the factors that influence evaluations as you design the evaluative process. Each evaluation is unique. One size does not fit all. Once all the process questions have been satisfied, it is likely a useful evaluation will result.

7.8 Case Study: Evaluation. Global Education Project in Central Asia

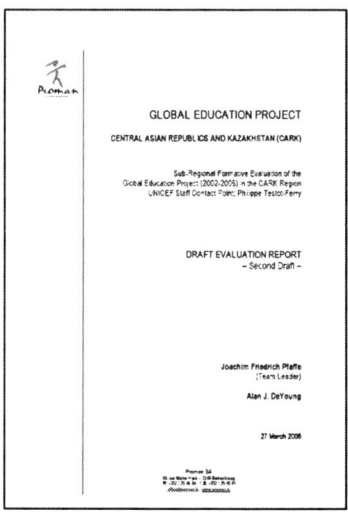

UNICEF (the United Nations Children's Fund) sponsors and operates several hundred education, environment and health projects around the world at any given time. To keep tabs on its global dedication of resources, UNICEF has rigorous evaluation requirements (UNICEF 2004). All projects must adhere to these standards of reporting or face stoppage of funding. In the former-Soviet countries of Central Asia, UNICEF supported school modernization through a Global Education Project, from 2002 to 2007. The project's mid-point evaluation report uses a wealth of qualitative and quantitative data to explain the unfolding of an ambitious and apparently underfunded project, seeking to infuse reform into entrenched educational bureaucracies (Pfaffe and DeYoung 2006). Methods used are structured interviews with opinion leaders; focus groups of key informants; field observations (called 'site visits' here); surveys of school teachers, administrators, students, and parents; and a secondary analysis (called 'desk review' here). Accessing sample groups in the four project countries was often difficult; only a handful of questionnaires from Tajikistan were completed, for instance, due to lingering unrest of civil war. Nonetheless, the evaluators apply their body of evidence to make far-reaching recommendations for the future of the project. Semantically, they suggest the term 'global education' be replaced with 'child-friendly schools.' Strategically, they suggest the whole project be absorbed within larger school reform plans, focused on national policy changes and on-the-ground teacher training. That these conclusions are stated openly shows UNICEF was open to hearing an accurate description of the project's merit and impact and was willing to accept failure as an option. The evaluation was conducted by an outside consultancy and cost €89,344.

References and Further Reading

Altschuld JW, Witkin BR (1999) From needs assessment to action: transforming needs into action strategies. Altamira Press, Lanham, MD
Bennett C (1976) Analyzing impacts of extension programs. Sage, Newbury Park, CA
Campbell D, Stanley JC (1963) Experimental and quasi-experimental designs for research. Rand McNally, Chicago
Creswell JW (2008) Research design: Qualitative, quantitative, and mixed methods approaches, 3rd edn. Sage, Thousand Oaks, CA
Diehl M, Stroebe W (1991) Productivity loss in idea generating groups: tracking down the blocking effect. J Personality Soc Psych 61:392–403
Dillman JA, Smyth JD, Christian LM (2008) Internet, mail, and mixed-mode surveys: the tailored design method, 3rd edn. Wiley, Hoboken, NJ
Donaldson SI, Scriven M (eds) (2002) Evaluating social programs and problems: visions for the new millennium. Claremont Symposium on Appl Soc Psych, 2001. Lawrence Erlbaum, Mahwah
Fink A, Bourque LB, Fielder EP, Frey JH, Oishi SM, Litwin MS (1995) The complete survey kit, vols 1–9. Sage, Thousand Oaks
Fink AG (2002) The survey kit, 2nd edn. Sage, Thousand Oaks, CA
Fitzpatrick JL, Sanders JR, Worthen BR (2003) Program evaluation: alternative approaches and practical guidelines, 3rd edn. Allyn & Bacon, Boston
Hansen HF (2005) Choosing evaluation models: a discussion on evaluation design. Evaluation 11(4):447–462
Hox J, Leeuw ED, Dillman DA (eds) (2008) International handbook of survey methodology. Lawrence Erlbaum, New York
Isaac S, Michael W (1995) Handbook in research and evaluation: A collection of principles, methods, and strategies useful in the planning, design, and evaluation of studies in education and the behavioral sciences. Edits Publishers, San Diego, CA
Miller DC, Salkind NJ (2002) Handbook of research design and social measurement, 6th edn. Sage, Thousand Oaks, CA
Patton MQ (2008) Utilization-focused evaluation, 4th edn. Sage, Thousand Oaks, CA
Pfaffe JF, DeYoung AJ (2006) Global education project, central Asian republics and Kazakhstan (CARK), sub-regional formative evaluation of the global education project (2002–2005) in the

References and Further Reading

CARK region, UNICEF Staff Contact Point, Philippe Testot-Ferry, draft evaluation report, second draft. Proman SA, Bettembourg, Luxembourg

Royse D, Thyer BA, Padgett DK, Logan TK (2009) Program evaluation: an introduction, 5th edn. Brooks Cole, Belmont, CA

Scriven M (1991) Evaluation thesaurus. Sage, Newbury Park, CA

Soska T, Johnson Butterfield AK (2005) University-community partnerships: Universities in civic engagement. Routledge, Binghamton, NY

Stufflebeam DL, Shinkfield AJ (2007) Evaluation theory, models, and applications. Jossey-Bass, San Francisco

UNICEF (2004) UNICEF Evaluation Report Standards Evaluation Office. UNICEF, New York

Watson D (2007) Managing civic and community engagement. Open University Press, Maidenhead, UK

Weiss CH (1997) Evaluation. Prentice Hall, Englewood Cliffs, NJ

Wholey JS, Hatry HP, Newcomer KE (eds) (1994) Handbook of practical program evaluation. Jossey-Bass, San Francisco, CA

Chapter 8
Characterizing the Mass Media

8.1 Introduction

Communicators today have a bewildering array of media choices for transmitting messages. Even the tried-and-true term 'newspaper' might mean a city daily, a neighborhood or rural weekly, a national daily such as the *Wall Street Journal* or *USA Today*, an on-line edition of any of those, an on-line-only news source, an alternative weekly, or a school publication. Advances in technology have created opportunities for communication far surpassing those of just a decade ago. Too often, a mode of communication is chosen without analyzing whether that particular medium is best for achieving the communication's goal. If your planning process has been thorough, then choosing among media choices will be based on your goal and objectives, target audiences, message characteristics and budget. All of these are made explicit in a well-done communication plan. Do your job well in formulating the plan, and media selection will be smooth. Fail to write a sound plan, and you will be confused and dazed by the myriad of choices out there.

8.2 Convergence

Adding to the potential for confusion as well as the sheer fun of creative message placement, boundaries between media now overlap and intertwine. There are more media to choose from, even as messages move through multiple media simultaneously. The information environment is one of rapid flow, braided delivery channels and a strong possibility of flooding, like a mountain stream in spring.

The coming together of multiple streams of information is called 'media convergence.' Jenkins (2006) states this definition: 'the flow of content across multiple media platforms, the cooperation between multiple media industries, and the migratory behavior of media audiences who will go almost anywhere in search of the kinds of entertainment experiences they want.' He argues the roles of producers and consumers have merged into just one: participant. The new rules of media engagement, however, are by no means written yet.

Media convergence is made manifest by multi-function devices. If you have seen a mobile phone, which is also a still and video camera, which is also a music player, which is also Internet-capable, you have witnessed the stuff of convergence. Computer-based communication can hardly function without convergence, with Web surfing requiring all sorts of 'add-ins,' 'plug-ins,' 'extensions' and 'apps' to access the latest and greatest content. The Holy Grail of convergence is 'the black box,' a single device that does everything. But, Jenkins (2006) calls this dream, a 'fallacy.' Instead of a single black box, he sees many black boxes, quickly obsolete and with always more bells and whistles.

Convergence of content, on the other hand, happens all the time. Movies and TV shows can be delivered to your phone. The same popular songs from musical artists can be played on computers, stereos, and personal players. Readers of newspapers, magazines and books can chose to forego paper and read on a screen. When a person seeks content, and perhaps makes their own content also, they are participating in convergence. Far beyond technological advances, convergence is a progression of social behavior and interactions too. That a new media form of connecting is called 'social networking' bears this out.

To explain the effects of convergence on media industries, an artificial line can be drawn between 'old media' (traditional newspapers, printed magazines and books, over-the-air radio and TV, etc.) and 'new media' (digitally delivered publications and music, on-line forums, web logs, etc.) (Flew 2007, 2008). Predictions of the demise of all old media are exaggerated, since some old media have skillfully morphed into new media. Virtually all magazines and newspapers operate Web sites now. Take, for example, a feisty little newspaper named the *Canyon Country Zephyr,* the masthead of which reads, 'All the News that Causes Fits.' Started in Moab, Utah, USA, in 1989, it has provided an unflinching defense against crowds of tourists and development to cater to them ever since. Publisher Jim Stiles was a friend and coworker of the late environmental writer Edward Abbey, and does his best to carry on Abbey's cantankerousness. Content pushes debates about development and environmental ethics to a small, fiercely loyal audience. In 2009, the paper moved, grudgingly, to an all-digital format – 'Planet Earth Edition…free to almost 7 billion people via the world wide web' (Stiles 2009).

Contrast distribution patterns between old media – one-to-many with little feedback – and converged media – many-to-many, interspersed with one-to-one and one-to-many, with lots of feedback. Collaboration and empowerment are new hallmarks, as individuals add their own content, piggybacking messages and sending them on down the line. Large media companies have seen their monopoly on content broken, while attempting to transform their business models to stay viable in the converged marketplace. Hard hit have been medium-sized newspapers. On-line classified ad services, such as craigslist, took huge chunks of revenues away and a significant number of newspapers have folded. Meanwhile, converged and new media keep evolving and figuring out how to generate revenue with as much gusto as they do content.

Literary great O'Connor's (1965) last collection of stories, published posthumously in 1965, was titled *Everything That Rises Must Converge*. It was her synopsis and ode to the teachings of French philosopher and Jesuit priest de Chardin (2009), who sought to unify theology, philosophy and science, pointing toward a complete merging consciousness and complex orderings of matter. Whether or not we are moving in that direction remains to be seen. Still, O'Connor's phrase presciently describes today's blurred boundaries between message-makers, messages, media, technology, and message-receivers. Communicators now craft convergences.

8.3 Characteristics of Mass Media

In this chapter, we analyze communication media, characterizing them not by type: newspapers, magazine, newsletters, billboards, radio, television, interpretive talks, etc. With new media and convergence, using such a typology now strains usefulness. Distinctions among types have blurred. Computers connected to the Internet as well as multi-function devices pulling information off wireless networks have expanded the modes available dramatically. Is an on-line magazine more a magazine or a digital news service? Are advertising panels in shopping carts more like billboards or display ads in print media? Compiling an inclusive list of types of media is no longer possible.

Instead, we will look at communication media based on qualities that cut across these outdated classifications. These characteristics include a medium's purpose, audience focus and depth, delivery mechanism, timeliness, and cost. Each of these characteristics can be evaluated separately by a communicator when selecting channels for their environmental messages.

8.3.1 Purpose

It may sound cynical, but the main goal of most media is to make money. Most newspapers and magazines, for example, make their profits by selling advertising. Analyze your local newspaper by calculating the amount of space taken up by advertisements in comparison to the amount containing news. You might be amazed at the amount of space set aside for ads. A healthy newspaper is filled mostly with advertising and not news. Likewise, a viable news website will have noticeable pop-ups, banners and eye-catching animations. In order to sell this much advertising, news media need to be able to demonstrate that their product is consumed by audiences the advertisers wish to reach. Just like environmental communicators, advertisers have well-researched concepts of the groups of people with which they want to communicate. Audiences have specific characteristics and each news media outlet offers an unequaled group. For the largest of mass media – the television networks – viewership size and demographics are demonstrated during 'sweeps' weeks and through Nielsen ratings. Newspapers and magazines closely track their circulation and readership demographics. Internet content providers have rapidly gotten more sophisticated in data they provide about their audiences. Most news organizations work to sell their audience to advertisers at least as hard as they work to sell their product to consumers.

Even those media distributed by non-profit organizations, or that are free to the public, usually must make enough money to cover production costs. Public radio and public television must raise funds to cover those costs not subsidized by their corporate contributors and the government. Free publications usually come from corporations or non-profit groups that fund their communications through indirect means, such as advertising or sponsorship. Internet provider services usually have a price for subscription, are provided by a company or school, or send lots of advertising to their subscribers.

The financial goal inherent in most media is important to consider when analyzing whether they are appropriate for a specific communication program. For example, newspapers are interested in those stories that will sell more newspapers, and thus will be reluctant to cover a story that will be of interest to only a few readers, regardless of how important the communicator believes the story is. Likewise, television networks are unlikely to run a public service announcement for a non-profit group during a time when there is a high demand for advertising spots, such as during a popular reality series or sporting event.

While the main goal of most media is to make money, they generally have one of three other purposes: information, persuasion, or entertainment. These purposes are highly interrelated, and often difficult to distinguish from a cursory analysis. But, knowing this additional purpose can help determine whether the communication mode is appropriate for your message.

8.3.2 Providing Information

Many media provide information in the form of news. Not all information is news, however. News is information that is important, interesting and timely. Environmental communicators have access to news media in several ways. First, news releases are routinely used to announce activities of an organization, such as new hires, new initiatives and public events. Communicators can also help to create news in order to gain news coverage. Consider organizing a demonstration, putting on a sporting event as a fundraiser, sponsoring an expert or celebrity to speak, or inviting reporters to join your organization's workers as they accomplish an important task. A third means of making your work into news is to provide information that helps the news outlet do its job better. Offer yourself or an expert from your organization for an interview on a pertinent topic. Send your area reporters a packet of background information describing what you do. Write letters to key reporters in your community explaining what you do that might be newsworthy. For each of these approaches to be effective, the communicator must have a thorough understanding of the news process, and be able to frame information in a way that is interesting to the reporter, editor and final reader.

Other media are used to provide information. For example, scientific journals are highly specified publications where scientists report findings of their research. Many organizations publish various documents intended to provide information on specific issues or events. One example, the fact sheet, is a self-contained document that provides background information in an easily understandable format. Fact sheets can be found all over the place, including all across the Internet.

8.3.3 Persuasion

Media meant to persuade range from speeches of those running for political office, to World Wide Web sites posted by advocacy groups, to handwritten opinion letters. All persuasive communication tries to get audiences to undertake a specific action.

Opinion letters are perhaps the most common example of persuasion. Generally, they are a personal appeal sent to an elected official, an agency official or a business in order to achieve some desired effect. Letter writers can be highly effective in influencing legislation and rulemaking, especially when combined with other writers in a well-focused campaign. Other media whose purpose is persuasion include advocacy magazines, billboard and placard advertisements, public service announcements, and newspaper editorials.

8.3.4 Entertainment

A large number of media are concerned primarily with entertaining their audiences. While these media may appear useless for communicating environmental messages, creative communicators can use entertainment to achieve their goals. For example, the Earth Communications Office works behind the scenes in Hollywood to infuse environmental messages into television series episodes and movies. Their public service announcements have been spliced onto feature films and distributed throughout the world. (See examples of their work at www.oneearth.org.)

Environmental interpretation often contains a large dose of entertainment in order to attract non-formal audiences. In addition, movies, television programs, cartoon books and music can provide important messages in support of environmental communication campaigns. For lower prices, items such as T-shirts, bumper stickers and novelty items can be employed to achieve communication plan goals. Entertainment is a powerful tool when employed creatively.

Communicators have differential access to different media. For example, environmental communicators have limited access to feature newspaper articles in that a reporter and editor have control over what gets written, where the article is placed, and when it appears. Democratization – less control and more access – is one of the beauty's of the Internet. When an environmental organization stakes out a Universal Resource Locator (their 'web address') and puts up a site, they are fully in control of content.

8.3.5 Audience Focus and Depth

When choosing among communication modes, communicators must carefully match their target audiences with the stated audiences of candidate media. Some media, such as newspapers, have wide, general interest audiences defined more by geography than any other factor. Alternately, many insider newsletters have small, diffuse, narrowly focused audiences. No medium reaches everyone with a particular set of demographics.

An example of highly specialized media is scientific journals. They can be important sources for environmental communicators seeking new knowledge working its way through the knowledge filter. In scientific journals, articles are peer-reviewed, meaning that scientists in the field review all articles submitted to a journal, make suggestions for improvement, and accept only those articles that are deemed to be exhibits of high-quality research, meaningful findings, and of interest to the journal's readership. Audiences for scientific journals are almost entirely made up of scientists working in the discipline covered by the journal. Some highly respected journals, such as the *New England Journal of Medicine*, *Nature*, and *Science* are read by wider, albeit technically trained audiences, including practitioners and reporters from mass media outlets.

Access to scientific journals is through direct submission by author-researchers. A peer review process generally involves one to four reviewers who evaluate the article without knowing the identity of the author or authors. All articles published in scientific journal require an abstract, and some journals publish only abstracts. An abstract is a short summary of an article, document, talk, or presentation.

An example of media with wide audience focus is news magazines, such as *Time* and *Newsweek*. Millions of people, from all walks of life, interests, and demographics regularly read these magazines. While they may seem out of reach for many environmental communication campaigns, large conservation organizations often get their issues covered by these magazines. Recently, media mogul Ted Turner was on the cover of *Time*. The photograph showed him wearing a tie with The Nature Conservancy logo. That's good publicity from an article not focused on conservation.

It is important to be careful in analyzing audiences for media. The World Wide Web may seem like a widely focused medium with the potential to target millions of computer users. But, because of the way users search for web sites that interest them, even sites that appear to have thousands of visitors may have little holding power, and thus actually have only a small, narrow audience.

Questions of holding power get at the loyalty and intensity with which audience members interact with any particular medium. Media convergence has opened wide horizons for interactivity between messengers, messages and audience members. The prevalence of opportunities to comment, post, upload, and talk back is astounding. And, the sheer numbers of people using these opportunities is equally astounding. Voting on reality TV shows' various popularity contests and voting in national elections total comparable tallies.

Knowing the audience focus of media is critical to implementing a communication plan. Careful audience targeting of messages is achieved through careful choice of communication mode.

8.3.6 Delivery Channel

The delivery channel is the actual transmission form used. Traditional newspapers are delivered on paper, while radio broadcasts use airwaves. In the past, media were often categorized as either print media or electronic media. Today, however, this distinction is blurred. Magazines and journals are now found on the Internet, as are live news broadcasts. Compact discs include written text as well as sounds and motion pictures. Even more of the music industry has moved on-line, selling single songs instead of albums. Technology has greatly expanded the range of individual delivery mechanisms.

In analyzing delivery of a given message, the communicator first must consider appropriateness. For environmental messages, paper-based media may be inappropriate since many paper technologies are not sustainable and are wasteful. Some media may be considered unsightly. One may not want to run a beautification campaign with billboards.

At the same time, the communicator must consider other purposes for, and the long-term impacts of, various media. Paper-based media can be saved and used at later dates to document the communication. Electronic communications are often fugitive, though digital storage media are nearing the permanence of paper. People's ability to store and retrieve radio and TV broadcasts have widened in the last decade. Internet publishing software, such as Adobe Acrobat, now make 'publishing' something most anyone with a computer can do. Reprints from magazines, journals, and all sorts of agencies and non-profit organizations are a few clicks away.

In the field of environmental interpretation, the delivery mechanism can be as important as the message itself. Creative use of signage, slide programs, costumed characters and actual objects can ensure a message gets to the intended audience.

The delivery channels of various media should be considered during communication planning. In combination with audience focus and media purpose, the channel used can have an important impact on the success of a communication project.

8.3.7 Timeliness

Media have various time-oriented characteristics that are important in communication planning. These are lead time, frequency and longevity. Each of these impacts the delivery of a message to the target audience.

Lead time is the amount of time required to get the message to the audience. For example, a group that holds a public demonstration or a press conference may well be able to get their message on the evening news programs if television reporters cover the event. This same event might also be in the next morning's newspapers. While the demonstration or news conference itself may require many weeks of planning and organizing, once held the time to reach the audience is measured in hours. Again, the rise of the Internet as an information source has compressed lead times even more. Some events can be streamed live, others get posted to sites within minutes. Journalists now refer to the '24-h news cycle,' where items must be fresh all the time around the clock. Globally important events are reported in a matter of minutes on the Internet around the world. Environmental communicators are learning to work with these new characteristics, turning them into advantages and not constraints.

At the other end of the spectrum, some periodicals, particularly journals with heavy peer-review and narrow-focus magazines, may have lead times of a few months. Articles for magazines are often assigned to writers 6 months or more in advance. Communicators should consider these lead times as part of the implementation planning for their campaigns.

Frequency refers to the number of opportunities afforded by a particular medium for re-transmitting a message to a specific audience. This is not necessarily the same as how often the communication mode is distributed. A newspaper may be printed daily, but the communicator may be unable to get informational articles into the newspaper daily. On the other hand, advertisements can be purchased in daily editions of the newspaper for reinforcement of the message. Magazines are often kept for years after their delivery. Communication planners must determine to what extent message repetition, and the reinforcement that makes possible, will improve its comprehension by the target audience. Also, one must decide whether such repetition must come from the same communication mode. Frequency is important for initial impressions, especially those unplanned by recipients. When a receiver is pulling messages, however, powerful search engines and digital storage have downgraded the impact of frequency for many media. If something can be found without much hassle, it matters little how often it is transmitted.

Longevity is the amount of time the message will be available to the audience. Published materials generally have a lengthy longevity, especially if, like newspapers, magazines, and journals, they are retained in libraries and data warehouses available via Internet connection. Future accessibility can be an advantage when the message is one the communicator wants available in perpetuity. On the other hand, if a mistake is made, or if a past message is in conflict with a current one, this longevity can be a major drawback. Traditional electronic media had short longevity, unless someone in the audience recorded the radio or television broadcast in which the message was contained. Since around the turn of the millennium, this situation has changed. Many broadcasters now archive their material and make it readily available on-line. Email and World Wide Web pages may have short or long longevity, depending on whether the message is maintained or is deleted (and becomes what is called 'fugitive'). If you have experienced a '404 error: file not found' message on a Web browser, you know a broken link means the page that once was there has gone missing. Communicators must match their strategy to media with appropriate longevity for effectiveness.

8.3.8 Cost

Communication plans have financial budgets attached to them. This economic reality constrains media choices available for delivering a message. Costs can include direct purchase, such as the purchase of TV advertisement time, as well as cost of creation, such as the expense of hiring a public relations agency to create the advertisement. Likewise, construction of a web site has to be done by somebody with certain skills. An additional cost is staff time to manage the campaign. For example, a news release may be free to publish, but writing, editing and distributing it takes staff time that has a real economic cost to an agency.

As a rule, cost of media increases with size of the audience reached. So, newspapers with larger circulations have higher advertisement prices, likewise radio and television stations with larger audiences have higher advertisement prices. So far, Internet advertising rates have followed similar patterns. For direct communication modes, such as direct mailings, larger audiences have larger costs, although economies of scale can lower cost per piece mailed. Communication planners can, by dividing target audience size by total communication budget, calculate a cost per audience unit figure. Usually, media buyers compare this as a 'cost per thousand,' that is the cost per 1,000 members of the audience. Their abbreviation for this measure is 'CPM.' It can be used to assess relative merit of different media within the total communication project.

Media costs also increase with elegance. For example, the cost of a simple fact sheet can increase dramatically as the number of colors in its design increases. Also, typesetting, the use of photographs and other graphics, paper weight and finish and printing process all impact the cost of a fact sheet. The important point for communicators is to determine how much elegance is required for the message to be received by the target audience. Additional colors in a fact sheet may or may not add to the impact of the message. For each communication mode chosen, the communication planner must determine how much to invest in the medium to achieve, not the best, but the appropriate delivery of the message.

8.4 Conclusion

When choosing a communication mode, the environmental communicator tries to find the most efficient way to transfer a message to the target audience. The effectiveness of the message will be higher when the purpose of the medium is consistent with the message content and the objectives of communicator. Additionally, the communicator can accurately target the audience by analyzing the audience focus of potential media, and choosing the best fit. Delivery channel, timeliness and cost constrain the possible choices for message transfer by excluding audiences or exceeding the means of the communicator. The choice of media is an iterative process within the planning process. In the next chapter we highlight some of the more common modes of media that can be used.

8.5 Case Study: Converged Media. 'Earth Song' by Michael Jackson

As a musical number, 'Earth Song' was written and composed by Michael Jackson. It was recorded in 1995, released on the 'HIStory: Past, Present and Future, Book 1' album and then as a single in November of that year (Earth Song 2009). One of several socially conscious songs of the late 'King of Pop,' the number is an epic ballad focused on animal welfare, human rights and environmental destruction. In a media crossover familiar since the 1980s, a music video – shot on four continents – came out in 1996 and earned a Grammy nomination the next year. Leap forward a decade, 'Earth Song' is available digitally for purchase from several on-line sources, such as iTunes and Amazon, either as a single or part of the album, as is all of Michael Jackson's catalog. Fans post content, either appropriated from broadcasts and commercial digital formats or created by individuals, to several web sites with upload features. On YouTube, the leading video sharing site, 'Earth Song' is posted no fewer than 70 times. Videos can be found on at least eight other sites; lyrics, on at least a dozen more. With Jackson's death in June 2009, fan-generated content escalates and, like the song itself, builds through a righteous crescendo. Statistics for the first ten 'Earth Song' videos available on YouTube (from a search conducted July 15, 2009) demonstrate the culture of convergence. They have been viewed 23,471,017 times, with viewers leaving 55,782 ratings and 52,697 written comments.

Credit: Michael Jackson kisses one of the children on stage with him while singing 'The Earth Song.' Dave Hogan Getty Images Entertainment.

References and Further Reading

Beamish R (1995) Getting the word out in the fight to save the Earth. The Johns Hopkins University Press, Baltimore, MD
Calvert P (2000) The communicator's handbook: tools, techniques and technology, 4th edn. Maupin House, Gainesville, VL
Earth Song (2009) In: Wikipedia, the free encyclopedia. Retrieved July 14, 2009, from http://en.wikipedia.org/w/index.php?title=Earth_Song&oldid=302027075
Evans D, Bratton S (2008) Social media marketing: an hour a day. Sybex, Indianapolis, IN
Flew T (2007) Understanding global media. Palgrave Macmillan, Baskingstoke, UK
Flew T (2008) New media: an introduction, 3rd edn. Oxford University Press, Melbourne, Australia
Friedman SM, Dunwoody S, Rogers CL (eds) (1998) Scientists and journalists: reporting the science as news. American Association for the Advancement of Science, Washington D.C., pp 55–69
Hartmann T (2009) Media choice: a theoretical and empirical overview. Routledge, New York
Jenkins H (2006) Convergence culture: where old and new media collide. NYU Press, New York
Miller JD (1986) Reaching the attentive and interested publics for science. In: Friedman S, Dunwoody S, Rogers C (eds) Scientists and journalists: reporting science as news. Free Press, New York, pp 55–69
O'Connor F (1965) Everything that rises must converge. Farrar Straus and Giroux, New York
de Chardin PT (2009) In: Encyclopædia Britannica. Retrieved July 16, 2009, from: http://www.britannica.com/EBchecked/topic/585678/Pierre-Teilhard-de-Chardin
Parker LJ (2008) Environmental communication: messages, media & methods, 2nd edn. Kendall/Hunt, Dubuque, IA
Stevens RE, Loudon DL, Ruddick ME, Wrenn B (2005) The Marketing Research Guide. Routledge, NY
Stiles J (2009) Take it or leave it…. Canyon Country Zephyr, Moab, UT pp 2–3

Chapter 9
Highlighting Useful Media

9.1 Introduction

Media convergence mashes delivery systems, deliverers and their deliverables all together. What's an environmental communicator to do? Keep planning, keep the creative juices flowing, and keep on top of what your analyses, feedback and evaluations are telling you.

While a list of media is endless due to convergent combinations, the characteristics offered in the last chapter offer a means of picking and choosing those for your campaign. There remain a few old standbys that still come in handy and they faithfully provide fundamental building blocks for more sophisticated message-media. At the heart of all message-media packages is cogent writing and clear indications of what exactly you want the audience to do with your message. Conversely, low-quality writing and unclear asks can sink any campaign. With a well-stocked arsenal of literary skills, the rest will follow. In this chapter, we list such building-block media, grouped into traditional, new and converged categories. We also give suggestions on how to use them successfully.

- Traditional media
  - News releases
  - Letter writing
  - Abstracts and executive summaries
  - Public service announcements (PSAs)
  - Information sheets
  - Science writing
  - Direct mail
  - Newsletters
  - Interpretive talks/presentations
  - Films
- New media
  - World Wide Web (WWW) sites
  - Email

- Mobile device messaging
- Blogs (web logs)
- Social networking

• Converged media

  - Reader responses
  - Podcast tour guides
  - Viral marketing
  - Webinars
  - Streaming events

9.2 Traditional Media

These are the workhorses of communications. They've been around for scores of years, some more than a century. All require excellent writing skills, along with formatting, graphic design, production skills, and accurate lists of who to distribute them to. Their usefulness remains.

9.2.1 News Releases

A news release exists to inform journalists of events that reflect functions of organizations. These might be the election of a new president, the release of a new study, or an annual event pertinent to a community. Events will be of interest if they are timely and of broad interest to the news media audience. Occasionally, a minor crisis within an organization that affects a community may be best reported via a news release by the organization's staff writers who can ensure that the story is correctly given.

Mechanics of News Release Writing – A news release needs to be written on official organization letterhead. Also given in the heading should be a statement of when the information is pertinent for printing such as 'For immediate release' or 'Embargoed until [date and time].' A contact person with address, email and phone number is essential.

The release should open with a descriptive headline that concisely presents the major emphasis of the article. Then, the opening paragraph, or 'lead,' should cover the who, what, when, where, why and how of the event. The lead, presented in active voice, gets a reader's attention. The body of the release, written in inverted pyramid style (see 16.7), provides facts without editorializing. You cannot assume a reader has background information such as location, addresses, function of an organization, knowledge of abbreviations used within the organization etc. Proofread your work to ensure relevance to the audience, timeliness and use of correct grammar. These are critical to getting it accepted by the editor.

Common Mistakes in New Release Writing – Editors trash the majority of news releases they receive, because the releases:

- Aren't tied to a local situation
- Are not newsworthy

- Contain too much advertising fluff
- Are long and cumbersome
- Are no longer timely
- Contain redundant information
- Are poorly written and/or presented
- Contain a source which cannot be confirmed or information suspected to be too biased

9.2.2 Letter Writing

After many centuries, one of the most persuasive and powerful forms of communication from an individual is writing a letter. The advent of the digital age has done little to make the humble personally written letter obsolete. Indeed, an opposite effect seems to be happening. Electronic personal communications media (e.g., email, instant messages and text messages) is the most commonplace mode to communicate speedily. Volume of these messages is approaching 100 billion per day globally. Nevertheless, digital messages have a cold impersonal feel to them. Even with the development of the Internet's own slang, array of emoticons (symbols included to express emotions), and multitude of chatty abbreviations (e.g., LOL, BTW, IMHO, etc.), electronic messages are designed for fast, succinct transfers of information devoid of most of the richness of face-to-face communication.

To really make an impression that time and effort went into sending a message or opinion, the clearly hand-written or typed letter is still a top choice. The personal letter retains warmth and attentiveness.

Not all letters need be letters of complaint focused on a problem or issue. Some can be affirmative letters of support, praise and gratitude. Whatever your opinion, unless you write and tell a specific someone about it, you will never be heard and the potential recipient will never be able to appreciate it or act upon it. Before you write, however, do your homework. Know what specific action or situation you are asking the letter recipient to act upon. Or, if a product is the focus, define it specifically. Know what you intend write about before you actually do the writing.

Why Do They Remain Impressive? – A personal letter of opinion nearly always has an immediate effect on the recipient. This is simply because most people do not actually express their opinions in writing. Most people will readily voice an opinion, many will consider posting opinions to online forums (especially if they can retain their anonymity), but few will actually take time to write it down and send it off. While estimates vary widely, it can be assumed that each letter that is written and sent also corresponds to at least 100 peoples' opinions that were never sent.

Letters to Elected Officials – An opinion letter does much more than just express your opinion, it acts as a barometer of public opinion in general for the recipient. Of the different levels of governmental structure such as village, township, town, city, borough, county, state, federal, know which level is best to receive your letter. States and nations are often subdivided further into administrative, judicial and other specific jurisdictions. All levels are subject to various kinds of bureaucracies. Despite the skepticism given governmental officials, it helps your write a better letter if you assume they really want to do the right thing as determined by their constituents. To this end they need to hear from constituents about their concerns. As a tip, write to the lowest level of official that can deal with your concern before moving higher.

Levels of Rule-Making – There are many levels of government active in legislation, regulation and other rule-making. Target the correct level where you want your letter to be read. You need to match the jurisdiction of the authority with the topic of your opinion. Who has the authority or capability to act on your opinion? It is also important to act through proper channels of a chain of command. Begin with the lowest appropriate level of authority able to deal with your letter's content. If you get no action or satisfaction at that level, write to higher authorities, and be sure to detail the lack of action from the lower parts of the chain.

Letters to Editors – Letters to the editor are another major option for expressing your opinion. The target audience for these is not really the editor, but rather the readership of a particular printed medium. Newspaper and major news magazine editorial letter sections are pulses of public opinion. Hence, legislators and policy-makers will also get a sense of the prevailing winds of public opinion from these sources. Use of this channel may also prompt more people like yourself to act on an issue or appreciate something of value you have expressed.

Letters to Manufacturers and Other Businesses – These types of letters let manufacturers and businesses know where they are, or are not, succeeding. Their success depends entirely on good consumer reaction to their products or services. Any poor reactions can result in much negative response being spread by dissatisfied consumers, while positive reactions can help promote positive responses. It behooves an organization to maintain good relations with its consumers and to mitigate and remedy problems before they become damaging. This is also the essence of public relations. Your opinion letters of affirmation or complaint are a primary source of feedback for organizations that offer a product.

Business Writing (Memos, letters, policy papers etc.) – Business writing, like any other, demands succinctness. Try to place yourself in the position of the message recipient. Would you want to read this communication if you were busy?

Before sending a letter, ask yourself:

1. Is this communication necessary?
2. What is the purpose of this communication?
3. Who is my reader? (If there is no audience, why are you sending it out?)
4. What are the relevant points to cover?
5. How would I like for my reader to respond?

There are five basic ways to sequence content for business correspondence: chronologically; by priority; by problem, cause, and solution; using comparisons and contrasts; with advantages versus disadvantages. Whichever sequence you use, keep the letter short and to the point. That will go a long way in ensuring it gets read.

9.2.3 Abstracts and Executive Summaries

In today's fast-paced and media rich society, most people are overwhelmed with too much information. Any lengthy report or scientific paper is expected to have attached a summary. Usually these pieces appear at or toward the front of a document. For reports, the summary is often referred to as an 'executive summary.' For scholarly works, the preferred term is 'abstract.' An executive summary or abstract need be complete enough to give the reader an accurate overview of the material contained in the larger document. This synopsis should also entice the reader to want to read further into the material. Obviously when time is a constraint, as it is for many professionals, the abstract or executive summary may be the only part of the document that actually gets read. Therefore, it behooves a communicator to carefully write this piece to ensure that the material is considered at all.

What Is an Abstract? – It is an essential condensation of the major points. It is non-evaluative, providing a succinct overview of the contents. The purpose is to provide a reader with enough information to determine whether the abstracted document is relevant enough to read in its entirety. So, the abstract is a time-saving device. It does not reflect the opinion of anyone other than the author of the original communication.

Mechanics of Writing Abstracts – A good abstract is accurate, self-contained, concise and specific, non-evaluative, coherent and readable. It should open with a lead sentence that sets the stage and explains the topic of the original article. The body should contain no more than a couple of hundred words, preferably on one page, in which all of the major concepts of the main material are brought to a reader's attention. It should end with a one-sentence conclusion that pulls the abstract together and helps a reader evaluate the original document's worth.

9.2.4 Public Service Announcements

Radio is a great medium to augment an ongoing communication program. One easy mode of using radio is with a public service announcement (PSA). PSAs are short announcements aired throughout the day using odd time slots during regular programming. Non-profit groups and agencies are the users since these slots are free. In the United States, the Federal Communications Commission requires radio stations to give air time in return for a license. But, the stations can choose what they air and when.

Radio stations are keen to air messages that are noteworthy, promote a cause or service they consider important, or inform the listening audience of something special. It behooves you, as a communicator, to get to know station managers to find common interests between the station and your organization. This then allows you to tailor your messages, so your PSAs appeal to a particular station and its listening audience. You might send in various lengths of your script so can be played in those odd time slots that unexpectedly crop up. Suggested lengths would be 15, 30, and 60 s long scripts.

In submitting a PSA, you may also send in a taped version along with the script, but usually the station will use one of its own announcers will read your script. Grammar rules for PSAs are relaxed. You are writing to be heard and must quickly create an image in listeners' minds. You can even use fragmented sentences, as they are often found in conversational speech. To ensure that your message is announced clearly and correctly, follow these tips.

- Plan on writing a conversationally paced script. A guideline is to have about 25 words per 10 seconds of air time.
- When you need a pause use three dots (an ellipsis), or more if a longer pause is required. This is preferred over semicolons and colons. Use a dash (a double hyphen) to indicate an abrupt change.
- Use an identifying phrase or explanatory title if beginning the message with an unknown name.
- Difficult names or words must be followed by a phonetic spelling in parentheses. For example, with the African name Thato Nadayitwayeko, the script will include: (Ta-Toe Na-Day-It-Waa-Yea-Koe). This makes pronunciation easier. Do not use parentheses otherwise. If an unusual name or word can be pronounced different ways then use phonetic spelling or underline the syllables where the accent falls. For example, the town of Waukesha would be pronounced correctly as (Wau-Keh-Shaw) or Wau<u>ke</u>sha and not Wau-kes-ha.
- Do not use abbreviations unless they are everyday ones. If you do, then use hyphens between each letter to let the announcer know each letter is to be pronounced separately. For example, U-S-D-A prevents the announcer saying 'us-dah.' And, not everyone knows what USDA stands for; it is officially the United States Department of Agriculture.
- Do not abbreviate the names of places or address identifiers such as street, avenue or boulevard. Neither abbreviate calendar terms such as days of the week and month, nor the titles of officials. However, you may use 'Dr.', 'Mr.', 'Mrs.', and 'Ms.' with a name.
- Use 'st,' 'nd,' 'rd,' and 'th' after days of the month, such as February 1st, April 2nd, July 3rd and October 10th.
- Phone numbers can be written as you would normally write them with hyphens, e.g. 999–555–1212. All other numbers should be numerical to 999, but then alphabetically written for the zeros. For example, 89 and 247 are fine, then write '9 thousand,' '21 million,' '67 billion,' or even '25.6 trillion.' Similarly when using fraction such as ¾ write out 'three-quarters.'
- Avoid sibilants – the 'S' words – that cause hisses and whooshes on the microphone. Also avoid using alliterations – tongue twisters – not everyone can say them easily on a first reading.
- If you have an address or a phone number in your message, have it repeated two or three times at the end so listeners will more likely remember it, or use an easy to remember mnemonic.

- When finished, have a colleague speak the script out aloud in the timed length you have specified. This will help catch errors and other problems an announcer may come across.

9.2.5 Information Sheets

Information sheets convey succinct amounts of information and are not meant to be comprehensive texts on a topic. They come in a vast array of sizes and formats. These can range from a simple one-sided fact sheet to a multi-page brochure. Whatever format is used, some simple features will make it attractive, the information appealing to read, and importantly, ensure the target audience picks up the material and reads it. Regardless of the type of information sheet used, the reader should be left with some form of action to pursue. This might be to phone or email a contact source for more information, or at least be given a series of references or further reading to find out more about the topic. Inclusion of WWW addresses will also lead readers to more sources of information. Not giving any outlets for more information will leave a reader frustrated if they want more information. Make it easy for them to follow an interest you may have generated. Consider not only the information, but how the target audience will obtain and use this information.

General considerations for any information sheet include:

- **Format** – Deciding whether to use a simple fact sheet, a folded brochure or a multi-page booklet, as just three examples, depends on the amount of the information needed, as indicated by audience analysis and agency objectives. Assess how the target audience will use this information and under what conditions they will receive it. For example, an informative fact sheet to be mailed to home-owners needs to be suitable for mailing. It also needs to be concise so it will be read, and light enough so postage costs are minimized. A three-panel pamphlet would be suitable to carry in a back pocket for someone on a short self-guided nature trail. A multi-page brochure would be useful to someone spending a few days exploring the flora and fauna of a wilderness refuge. However, backpackers intent on keeping their packs light will forgo a bulky brochure regardless of how informative it may be.
- **Color** – Decide if the sheet will be a single color type on a plain background, or many colors. Printing costs can be prohibitive for color documents, even if you plan on printing in-house. If the budget is tight, then monochrome printing may be the best option. Even basic black print on white background can be appealing if layout considerations are maximized.
- **Layout** – The title should be bold and stand out clearly from the rest of the text. It should also convey at a glance exactly what the information sheet is about. Breaking the page into columns or panels makes the information easier to read and gives the impression of 'sound bites' rather than diatribes on the topic. The font chosen needs to be easy on the eyes, large enough to read, yet allow the inclusion of suitable amounts of information. Graphics should be clear and easy-to-understand.

Each graphic should be closely associated with the corresponding text in the layout. Alignment of text, margins, borders, shading, graphics, section separators and judicious use of 'white space' all support an appealing layout.

9.2.6 Science Writing

Scientific journals are not approachable to those without specialized training. Even scientists in one discipline can find a journal from another to be bewildering, the language foreign. Non-scientists can hardly be expected to ever want to look at a scientific journal. So, taking scientific findings and translating them for broader audiences is a highly marketable skill and one vital if society is to have broad scientific literacy.

A bit less than half of the American public pays attention to scientific news (Miller 1986). Many of these interested persons actively seek news of scientific discovery. Writing stories for them takes special skills. Honing an ability to construct stories from science is the bailiwick of science journalists. Gleaning story ideas from primary sources, the scientific journals, science writers then interview scientists in a manner that respects the tentativeness that is hallmark in the culture of science. Scientists as sources for news walk a tough line between public service and protection of hard-earned reputations. Journalists sometimes find scientists difficult and unrealistic in their expectations of the news media's role in society. Both scientists and journalists have strived to develop better relations by focusing on their common ground: the diffusion of knowledge.

The best science-focused news results in wider participation in policy-making, as informed citizens express their opinions as to how scientific findings can be incorporated into society's workings. Intermediaries can play catalytic roles in the generation and dissemination of important science news. Examples of such intermediaries include university public relations personnel, government public affairs workers, non-governmental activists, and many other professionals who have interest in spreading findings from research.

9.2.7 Direct Mail

A frequent, though diminishing, method for targeting specific audiences is direct mail. It can be as simple as a post card. If the post card is well-designed and targeted, it can elicit a desired response. Thus, direct mail is a technique rather than a specific medium. In the persistent age of mass mailings, it is easy to think of this format as junk mail, but it can be efficient and cost effective. So much depends on the mailing list used, as direct mail targets a segment of an audience of interest with a message pertinent to their needs. It also lets the audience follow-up easily with contributions to a cause, requests for more information about an issue, inquiries about products etc. In many ways, direct mail is a form of social marketing (see Chapter 19). Effective direct mail demands credibility, which includes a viable cause to support, products that appeal to the audience and match their needs. Accurate mailing lists match message to audience.

Some examples of direct mail are:

- Non-profit groups who solicit contributions to maintain their advocacy and lobbying efforts.
- Alerts from activist groups about arising situations. These types of messages can be easily personalized to make the audience feel appreciated.
- Special events, such as an Earth Day rally or a community clean-up day, can be promoted.
- Updating a target audience about a developing situation.
- Marketing ecological safe products via mail order to a geographically broad audience.

9.2.8 Newsletters

Rare is the environmental group which lacks a newsletter. Newsletters are regularly distributed publications which come directly from an NGO, agency or other well-defined group to its members, constituents or customers. Typically, newsletters are printed and folded to read as standard-sized paper (A4 or 8 ½ × 11 in.), with 4–16 pages and a quarterly, bimonthly or monthly frequency. As a direct conduit for a group to tell its own story, nothing beats a newsletter. Purposes for newsletters are to benefit people who have identified with your group, to give them something in exchange for their financial support, to reinforce the group's mission and accomplishments, and to expand understanding about the group's work (Agre 2007). Distribution of newsletters should take into account the preference of the recipient. If someone wants a paper copy, they should get that from the group. If an email version is OK, that probably saves resources for the group. Agre (2007) noted, 'most donors still like to get a paper newsletter. Many people do not like to read things on a computer screen, and you just can't reach everyone via email.' Notwithstanding, newsletters are usual and important components of web sites.

9.2.9 Interpretive Talks/Presentations

When a person talks in an organized fashion before an audience, a presentation is being given. From an almost impromptu discussion to a highly anticipated, ticketed speech with extensive high-tech visual aids given in a large venue by a recognized luminary, presentations take many forms. The Definitions Project (2007), an effort by 30 North American agencies and organizations to establish a common vocabulary among environmental interpreter and educators, states an 'interpretive program' to be: 'Activities, presentations, publications, audio-visual, media, signs, and exhibits that convey key heritage resource messages to audiences.' In presenting resource messages to audiences, via talks and presentations, an environmental communicator can seek inspiration in Tilden's (1957) original six 'Principles of Interpretation':

- Any interpretation that does not somehow relate what is being displayed or described to something within the personality or experience of the visitor will be sterile.
- Information, as such is not interpretation. Interpretation is revelation based upon information. But they are entirely different things. However, all interpretation includes information.
- Interpretation is an art, which combine many arts, whether the materials presented are scientific, historical or architectural. Any art is in some degree teachable.
- The chief aim of interpretation is not instruction, but provocation.
- Interpretation should aim to present a whole rather than a part, and must address itself to the whole person rather than any phase.
- Interpretation addressed to children (say, up to the age of 12) should not be a dilution of the presentation to adults, but should follow a fundamentally different approach. To be at its best it will require a separate program.

9.2.10 Films

Films – used interchangeably with 'motion pictures' and 'movies' – are well into their second century as a means of mass communication. Movies are a medium of intense power to evoke emotion and transport people across place and time. They deliver entertainment first and foremost. Information and education are also communicative purposes of this medium. Most big-budget blockbusters are produced in Hollywood, the American film capital, and Bollywood, the Indian industry hub around Mumbai. Technological advances have decentralized filmmaking.

Environmental messages deployed using video and film are within the reach of organizations, activists and artists with only basic equipment. Even a mobile device is now capable of serving as a movie camera. Slightly higher production values can be had for much less investment than a generation ago.

Murray and Heumann (2009) note green themes have appeared, often in connection with dark visions of apocalyptic futures, in such classics as *Soylent Green* (1973) and the *Planet of the Apes* series (1968–1973), as well as more recent works such as *Happy Feet* (2006), *The Day After Tomorrow* (2004), and *An Inconvenient Truth* (2006). A distinct genre of film devoted to environmental matters has spawned wildly popular environmental film festivals. Successful festivals have been held in the Canada, Czech Republic, Germany, Japan, Poland, Portugal, Russia, Slovakia, South Africa and the United States. Many festivals have open submissions, so a film-making environmental communicator may attempt to expand their audience and gain the boost that comes from juried acceptance to a reputable film festival.

9.3 New Media

The Internet and mobile phones have opened new worlds of electronic options for communicators. Communications through the digital platforms of computers and mobile devices is new media. Access to new media requires electronic machines – personal computers and mobile phones, for the most part. More than 1.6 billion people use the Internet (American Micro Devices Inc. 2009). And, there are more than four billion mobile phones in use globally (International Telecommunications Union 2009). New media are the standard-bearers of the Digital Age of communication. Though the beginning of the Digital Age defies being pinpointed, more and more information was moving on-line through the 1990s. By 2010, there exists a whole generation of teenagers and young adults who do not remember communicating without new media. They are the 'digital natives,' and their interactions reflect comfort and competence as new media communicators (Prensky 2001).

9.3.1 The World Wide Web

The World Wide Web (WWW) fast became the main means of new media connection and access to new media content, following its public free-access debut in 1993. WWW is the primary means for connected groups to find information and to present information. Web basics to be mindful of are:

- The design of a web page is crucial. The opening page must contain all relevant information minimizing a need to scroll. Links to other parts of the website must be visible and functional as well, with enticing captions. Page design is open to creativity, but should not distract from the message. The main drawback is competition, as tens of thousands of new pages are added every day. To be successful, a web site must be found. This means publicizing the URL ('web address') and maximizing search engine results.
- There is less consistent organization on the WWW as there is in a bricks-and-mortar library. Finding information you want can be time consuming and hit-or-miss. Getting to a useful site can be purely overwhelming, with the amount of information you do find, whether relying on search engines or links from sites you know well. Search engines do help you find information, though they use unique search strategies to locate key terms within a web document. Different engines will produce difference results. Becoming familiar with web browsers and search engines can help a lot, and by using several different search engines, you can learn to locate more obscure pages. Using different search strategies of your own to thin down the search selection field will also help in attaining a manageable search list to review.
- When a web page author develops a web site and then posts it to the WWW, it must be understood that unless people have that page's unique web site address they may not be able to find it through a search. Search engine companies have

'netbots' (automatic search programs) that continually scan web documents for key terms. Key terms contained in a coded line in the 'meta-data' headings are more easily found by these netbots. This however, can take weeks to months, if the document is ever found at all. A whole industry has sprung up to maximize search capabilities. Carefully select key terms to make your page rise above the many millions of other pages! Search assistance companies submit your website to search engines and also implants techniques to make the pages appear at the front of any search list when your key terms are entered.
- Validating information is essential. A beauty and a beast of the Internet is the lack of gate-keeping and peer-review of information placed on the WWW. An old saw from consumerism is apt here: 'Caveat emptor' ('Buyer beware'). Just because information is attractively appealing from a impressively named web site does not give it validity.
- Enormous amounts of information can be found on the WWW. Some is good, some not so good. Some is top-notch research data. Some is just unsubstantiated drivel. Some is placed to provide useful and useable information. Some is for persuasion. Some is parody and meant to be subversive. So, what can we do to judge content quality?

Evaluating a site means using your trained judgment to determine the veracity of the content. There can be good information on 'bad' sites and bad information on 'good' sites. Some can even be blatantly false. Information on the Web calls for an even more heightened application of healthy skepticism, compared to information from traditional mass media sources. Apply these questions to environmental information on the WWW (EETAP 1999):

- **Authority** – Who wrote the information? Are they credible? Where was their information derived from?
- **Audience** – Who is the intended audience? What is the focus of the web site?
- **Context** – Why is this information being presented to this target audience? Is there any obvious bias? Is the information broad or deep? What links are there to other pages?
- **Accuracy** – Is the information 'sound'? How do you know? Are sources given that can be verified?
- **Currency** – Is the information recent? Is it updated as appropriate?

9.3.2 Email

This mode of communication is fast and furious. More than 62 billion email messages are exchanged daily (Lyman and Varian 2003), making email the primary person-to-person means of exchanging messages quickly and easily, often across great distances. Email allows you to communicate with people worldwide with the push of a button. While email between friends may be more apt to ramble, business or non-personal email really needs to be to the point. People who receive extensive numbers of email messages a day soon stop reading them, or just scan the first few lines at best. In a sense, writing skills on email are even more necessary if you wish the recipient to read your transmission. While emoticons (such as ☺ or ☹) can help to add some personality to short messages, email remains a cold medium, and as such is subject to misinterpretation. People usually tend to send-off email messages without much editing. Sometimes, it is wise to let a message sit for a hour or two and then re-read it before sending. Much grief can be avoided by this simple practice. It has been said of more than one person in a senior position, 'they need a delay and re-read key, to keep their foot out of their mouth.' While it is a great way to reach a lot of people quickly, please do not intrude on people just because it is easy to do so. Be polite, reasonable and succinct!

9.3.3 Mobile Device Messaging

As home computers have shrunk in average size, from desk-hogging clunkers to today's thin and sleek notebooks, so have mobile phones evolved during the Digital Age. Mobile phones started out huge, requiring a shoulder strap to carry. They became smaller, but more importantly added communicative functions – text-messaging, picture and video recording, geographic locator capabilities, and Internet browsing. Multifunction mobile phones are also termed mobile devices.

Harnessing this technology for environmental communication can be done. For example, 45 agencies in greater London, Great Britain, collaborate on airTEXT, a service where air quality alerts are sent by text message (London Borough of Croydon 2007) to anyone who subscribes. The service is free and assists with the upkeep of public health.

9.3.4 Blogs

Blogs, a contraction of 'web logs,' are regularly updated sources of information posted by individuals to the Web. Bloggers are, by and large, citizen journalists who write about something they care about and want to tell others about. Some comment on big events of the day, while others are nothing more than electronic diaries of mundane daily lives. Bloggers in remote places cast light on happenings that may otherwise not be noticed. The most influential bloggers reach tens of millions, can affect mainstream media coverage, and may influence policy-making. Technorati (2008), a search engine, research and marketing firm catering to bloggers, says 'the ecosystem of interconnected communities of bloggers and readers at the convergence of journalism and conversation' constitutes the 'blogosphere.' Technorati's State of the Blogosphere 2008 report found 133 million blog entries in 81 different languages. Monthly visits per blog averages more than 20,000 and about half carry advertising of some form.

9.3.5 Social Networking

Interactions between people with common interests facilitated online have reached their most formal structures with social networks. Most social networks are Web-based applications which facilitate sharing of information and further communication. A core power of social networks is their leveraging of collective intelligence to improve the network itself. Flew (2008) noted these principles of on-line social networks:

- Many-to-many connectivity
- Decentralized control
- User-focused and easy to use
- Open technologically, so programming can be done by many
- 'Lightweight' in design, administrative requirements, and on-going maintenance
- Expected to evolve over time, as users modify them

Popular social networking sites are Facebook, MySpace, Twitter, LinkedIn, Hi5, Friendster and Xiaonei. Flew (2008) also includes Wikipedia, YouTube, and Flickr in his concept of social networking media.

Social networking represents the latest and greatest frontier of the new media landscape. Should you explore the territory and, perhaps, stake an environmental communications claim there for your campaign? Higman (2009) cautioned, 'You need to have your website, email marketing and online-fundraising ducks in a row prior to moving into the social-networking space.' Among the figurative 'ducks' she denoted are a website with up-to-date content (pictures, videos, podcasts and blogs), a functional means of collecting email contact information, and success with being found via search engines. Social networks are the most sophisticated media choice available right now and an organization had better be skilled in the building-block media and other new media before braving the social networks.

9.4 Converged Media

Immersion in mediated messages and unprecedented, widely available technological abilities to add messages to the global flow of information means convergence. How has media convergence affected the practice of environmental communication? Compare these two actual job listings:

> **1998 – Environmental/Energy Communication Specialist** Develop effective communication structure and processes, and cultivate greater visibility of the research activities and products at Idaho National Laboratory (U.S.). This includes, but is not limited to, such activities as creation of press releases, backgrounders, tip sheets to science writers, media alerts, newsletters, annual reports, creation and utilization of mailing lists of significant writers, establishing contacts with professional societies, creation of list of lab experts to be accessible to media, co-producing articles with scientists, participating in news conferences. Work closely with scientists, engineers and environmental technology managers to develop a 'research communication culture.'

Requires a degree in science or engineering, plus a minimum of 10 years of relevant experience in environmental communications. Candidates should have demonstrable experience in translating technical information into compelling prose, as well as helping to produce professional quality audio and video. Candidates should also have experience in helping the media obtain accurate, timely information for articles and news segments. Candidates should have substantial experience in translating technical information into lay language. Knowledge of, or experience with, Department of Energy is desirable. Salary is competitive.

2009 – Environmental Communication Specialist Can you use a video camera and produce short segments to educate and engage people? Have you produced podcasts and managed audio files? Are you familiar with pushing multimedia content out through online sites such as YouTube, Flickr and Facebook? Do you have graphic design skills?

If so, the Chesapeake Bay Program (Maryland, U.S.) is hiring a Multimedia Specialist to produce and manage a variety of content. Candidates should have 1–2 years of experience in video and podcast production, online content distribution and graphic design. The ability to use cameras, microphones and programs such as iMovie, Final Cut, GarageBand and Photoshop is preferred. Applicants should be highly organized, exceptionally creative, savvy with technology and self-motivated, as well as have a strong interest in environmental issues. This is a great opportunity for recent college graduates who have hands-on experience working with multimedia hardware and software.

For this environmental communications position, please email a resume, work samples and salary requirements.

What a difference 11 years makes. What a difference new and converged media make. Because convergence wreaks havoc on list-making, we will highlight but a few convergent media examples, with a few ideas of how they have been used by environmental communicators.

9.4.1 Reader Responses

Instantaneous and voluminous reader and viewer response has become such a part of Internet-based communicating that the sea change from traditional journalism's time-insensitive feedback loops can be forgotten. Pre-Digital Age Newspapers, TV and radio outlets may have felt they allowed adequate opportunities for feedback from their readers and viewers. Compared to the ample chances for response today, however, they were virtual one-way streets. Flow went out and, after a while, a trickle might have come back.

Today's community news outlets are ravenous for readers to respond to their content, and mightily encourage reader submissions. Within environmental journalism, consider *E/The Environmental Magazine*, a wide-circulating periodical founded in 1990 as a traditional paper-and-ink publication. It has survived, in part by adopting a dual-life – analog and digital. *E*'s digital form asks for feedback on every web page, through an 'Advice & Dissent Submission Form.'

Inviting participation is important, whether the situation is face-to-face or virtual. When seeking involvement on-line:

- Include easy-to-use response forms, preferably ones that collect respondent identifiers, contact data, and instantly displays their input.

- Moderate for vulgarity and off-subject content. Resist all urges to censor feedback you do not agree with.
- Use feedback as you would other evaluative data. Archive it for later use, both internally and to the wide world.

9.4.2 Podcast Tour Guides

Podcasts are easily produced audio- or video-installments which can be automatically received and synchronized with a mobile device so that listening and viewing occurs when the receiver wants it to (Mocigemba 2008). Another of the many portmanteaus of the Digital Age, podcasts take their name from Apple's iPod, one of the most popular mobile devices, and the last syllable of 'broadcast.' So, while a podcast will look and sound something like a traditional radio or TV show, they have a new capacity to cross time and have final delivery when desired by the end-user. No transmitters are needed. As converged medium, podcasts hit the eyes and ears like broadcasts of old, though they travel through the Internet as their delivery and feedback service.

Two examples of environmental communicating via podcasts are:

- EarthLink, in June and July 2008, posted a 'make advertising better' challenge. Produce a simple but effective advertisement for EarthLink and upload it. Each week during the contest they named a weekly winner (and gave them a $250 prize). At the close, it was a grand prize of $2,000. The ads were then used on many podcasts and other media.
- Museums have led the embrace of podcasts as tour guides, replacing paper brochures or special audio set-ups. Places where environmental interpretation happens have followed. Many national parks, nature centers, and historical sites now offer podcasted guides for visitors. For example, Grand Canyon National Park, Arizona, US, offers free ranger-produced podcasts of 'an insider's look' at the park, as well as hiking and river-running.

9.4.3 Viral Marketing

The new epitome of converged media splendor is when a message is able to 'go viral.' Viral marketing works when individuals pass along messages, perhaps adding their own short vote of confidence in it. Email forwards are one form. Like real viral pandemics, viral marketing successes show exponential growth and exploit the resources of larger entities. A large portion relies on humor. They flare seemingly out of nowhere and capture attentive wonderment, if only briefly. For environmental communicators, there are a few green viral marketing highlights. Blogger Brian Clark Howard (2008) listed 'The Web's Best Environmental Humor,' allowing you to scroll down a Top 10 funniest green videos. To vault *The Green Collar Economy* onto the New York Time best sellers list in its first week of release, first-time author and long-time activist Van Jones used a deceptively simple viral strategy: he called and emailed everyone he had contact information for and asked them to forward his book announcement to everyone they could (Shaloff 2008).

The fluency of digital natives will likely make them well-suited for designing viral messages. 'So we see a familiar pattern in the history of the Internet, where the use of the new medium for strategic communication, dialogue, conversation and community building co-exist with its uses for subversion, parody, invective and misinformation,' said Flew (2008). In order to become virulent, a package has to contain a core message, but be swaddled in some coating – be it satire, shock, irony, or just plain funny – that gets it through the usual defenses of recipients.

9.4.4 Webinars

Though stodgy in comparison to viral messages, webinars exploit efficiencies made available by the Internet. A webinar – '*Web*-based sem*inar* – is an live Internet-delivered presentation allowing people to participate via their computers, rather than coming together in a single location. There are two main types: webcasts (one-to-many) and web conferences (many-to-many, depending on the level of participation allowed).

Internet business consultancy Frost and Sullivan (2006) stated the main features for webinars:

- Pre-event features
    - Scheduling and setting up
    - Customized invitation, registration and reminder
- During-event features
    - Slide presentation
    - Desktop sharing (screen sharing)
    - Application sharing
    - Web co-browsing
    - File transfer
    - Internet audio and video streaming
    - White boarding and annotation
    - Chat (instant text messaging)
    - Polls and surveys
    - Recording
- Post-event features
    - Thank you and follow up emails
    - Report compilation and data analysis
    - Playback

Quite a few service providers have systems, usually available by subscription, which provide these features. At their best, webinars save resources by reducing travel while at the same time retaining much of the intimacy of a classroom setting for learning. Great webinars turn a computer-mediated lecture into an engaging conversation.

Here are titles of actual environmental communications webinars:

- 'Green Building Underwriting: Increasing Cash Flow & Reducing Expenses – Monetizing Greenhouse Gas Emission Reductions for Real Estate Transactions'.
- 'Making Environmental Education Relevant to Culturally Diverse Audiences'.
- 'Sustainable Seafood – Sustainable Seas'.

If the webinar facilitators have mastery of the technology, the learning experience will be enhanced, and the result, though not identical to that of as a single-location workshop or class, no less of an experiential learning episode.

9.4.5 Streaming Events

Like webinars, streaming events live (or in near real-time) on the Internet may have lost its novelty, even though doing so would have been a huge technological stunt less than 10 years ago. When media are streamed, they are constantly received at the end-users computer or mobile device. Streaming requires Internet connectivity, whereas traditional TV and radio broadcasting goes out over the airwaves. Environmental groups have streamed events to broaden their audiences and also to create lasting records of their work. Commonly saved streams include interviews with notable advocates and officials, available at the click of a link for viewing and listening again. A variation on this convergence theme is uploading video content to networking sites such as YouTube.

We wish to share two environmental organizations with many years of experience at streaming their events, the JASON Project and Bioneers.

JASON Project – Founded in 1989 by the discoverer of the Titanic wreckage Robert Ballard, this education program studies different compelling topics in science each year. JASON Project connects students with scientists using a variety of new media tools (The JASON Project 2007–2009). Since the early 1990s, streamed video conferences have been part of their communications mix. Students from a region gather a satellite site for the event. They are able to see and be seen by all the other satellite sites as well as the main 'mission center,' where scientists teach and conduct research.

Bioneers – Founded in 1990, this group through their annual conference studies natural medicine and bio-cultural diversity (Collective Heritage Institute 2009). Each October, about 3,000 attend the conference in the Bay Area of California, U.S. Early on, this environmental event pioneered the use of satellite and Internet to beam and stream portions of the conference to partner sites. For 2009, there were 18 'Beaming Bioneer' spin-off conferences.

9.5 Unusual Media for Environmental Communication

Most anything you can think of can be used to deliver a message: public hearings, conferences, action days, field trip, rallies, open houses, art contest, eco-tours, street fair, celebrity photo-ops, church programs, youth groups, benefit concerts, teacher workshops, demonstration, trail walks, puppet shows, T-shirts, balloons, bumper stickers, door hangers, utility bill inserts, parades, comic books, coloring books, pens, pencils, erasers, place mats, posters, exhibits, baseball hats, coffee mugs, toys, flags, songs, poems, museum programs, story books, billboards, murals, photographs, curriculum materials, software, electronic bulletins, slide shows, video games, cartoons, field guides, information kits, speakers bureaus, calendars, and pieces of music. The list goes on and on. All can become suitable channels for messages with the help of a competent environmental communicator. Alternative formats should be support media, not primary message conveyors. Use alternate means for simple messages in support of more complex communications via more conventional channels.

A communicator selects and designs the media that will best address the purpose of the message and reach their targeted audiences. Whereas selection of traditional media used to entail a one-time per campaign, 'little of this, a little of that' approach, new and converged media open opportunities for communicators to propagate messages and see if they will replicate throughout a target population. Campaigns still are planned with attention to cost, focus, and timeliness, but are no longer dictated to by the zero-sum shackles of press runs and broadcast air time. Environmental communicators can go viral, infiltrate social networks, and mix and shuffle their media choices.

9.6 Case Study: Useful Media. Plant a Billion Trees (http://www.plantabillion.org/)

This campaign to reforest the Atlantic forests of Brazil has a simple goal: raise $1 billion to plant 1 billion native trees in 7 years. Their convergence-rich web site includes intuitive navigation with fact sheets and education videos embedded. It has a running tally of the number of trees planted, providing an instant measure of progress. Visitors can pass the message along, with e-cards or even your own 'Plant a Billion' web page. Of course, there's a donation interface, where gifts from $1 to $1,000,000 can be made. Navigate a bit deeper into the site (which you will notice is actually part of parent organization The Nature Conservancy), and you might find T-shirts, a water supply game, content from Brazilian photographers and foresters native to the region, and magazine-style stories about the forest and its conservation (The Nature Conservancy 2008). The campaign is new media joy, with hypertext access, graphics often preceding text, and clear asks for participation.

Credit: Plant a Billion logo. The Nature Conservaney.

9.7 Case Study: Useful Media. Motorola Renew and Samsung Reclaim

Mobile device manufacturers took big steps toward a convergent environment-and-communications singularity in 2009. If media, message, and messenger ever have a perfect merging, they'll use phones something like the Motorola Renew or Samsung Reclaim. In February, Motorola unveiled its Renew model, available through the Best Buy chain with a T-Mobile contract. The model is certified as carbon-neutral in its manufacture. Its plastic is made from recycled water bottles. In August, Samsung countered with Reclaim. Service is attached to a Sprint contract, but all sales also prompt a donation to The Nature Conservancy. The phone is made of 40% corn-derived plastic and its charger is Energy Star rated. Features include one-button access to Facebook, Twitter, MySpace and YouTube. For a person with an environmental worldview, material possessions need to reflect their beliefs and provide them benefits of status in the eyes of their friends seeing them use such items. Perhaps the fabled 'black box' is really a green one.

Credit: Samsung Reclaim Logo. MWW Group.

References and Further Reading

Agre D (2007) Newsletters: an essential part of the fundraising mix. Grassroots Fundraising J 26(1):4–7

American Micro Devices Inc. (2009) World internet usage. http://www.50x15.com/en-us/internet_usage.aspx. Cited 8 Aug 2009

Beamish R (1995) Getting the word out in the fight to save the Earth. The Johns Hopkins University Press, Baltimore, MD

Calvert P (2000) The communicator's handbook: tools, techniques and technology, 4th edn. Maupin House Publishing, Gainesville, VL

Collective Heritage Institute (2009) http://www.bioneers.org/. Cited 9 Aug 2009

Definitions Project (2007) http://www.definitionsproject.com/definitions/index.cfm. Cited 5 Aug 2009

EETAP (1999) Evaluating the Content of Web Sites (1999) Environmental Education and Training Partnership (EETAP) Library, OSU Extension, 700 Ackerman Road, Suite 235, Columbus, OH 43202-1578

Evans D, Bratton S (2008) Social media marketing: an hour a day. Sybex, Indianapolis, IN

Fehrenbacher K (2009) 8 green cell phones: who's got 'em? earth2tech http://earth2tech.com/2009/08/06/green-phones-whos-got-em/. Cited 9 Aug 2009

Flew T (2008) Communication for the 21st century Or, how to have your blog and read it too! Paper presented to Society of Business Communicators, Brisbane, Queensland, 13 March 2008

Frost & Sullivan (2006) World web event services markets. Frost & Sullivan, Palo Alto, CA

Hartmann T (2009) Media choice: a theoretical and empirical overview. Routledge, New York

Higman RR (2009) 10 things you need to do prior to diving into social media. Network for Good Learning Center http://www.fundraising123.org. Cited 8 Aug 2009

Howard BC (2008) URTH Guy. The Daily Green. http://www.thedailygreen.com/living-green/blogs/recycling-design-technology/funniest-green-viral-videos-460808. Cited 8 Aug 2009

References and Further Reading

International Telecommunications Union (2009) Measuring the information society: the ICT development index. International Telecommunications Union, Geneva

London Borough of Croydon (2007) http://www.airtext.info/. Cited 8 Aug 2009

Lyman P, Varian HR (2003) How much information? 2003. http://www.sims.berkley.edu/how-much-info-2003 Cited 22 June 2009

Miller JD (1986) Reaching the attentive and interested publics for science. In: Friedman SM, Dunwoody S, Rogers CL (eds) Scientists and journalists: reporting the science as news. The Free Press, New York, pp 55–69

Mocigemba D (2008) P4P – podcasting for participation. Int J of Sustain Comm 2:3–21

Murray RL, Heumann JK (2009) Ecology and popular film: cinema on the edge. State University of New York Press, Albany

Parker LJ (2008) Environmental communication: messages, media & methods, 2nd edn. Kendall/Hunt Publishing Co, Dubuque, IA

Prensky M (2001) Digital natives, digital immigrants. On the Horizon 9(5):1–6

Shaloff N (2008) How environmental activist Van Jones' book The Green Economy reached the NYT best seller list. The Huffington Post, http://www.Huffingtonpost.com/ cited 8 August 2009

Stevens RE, Loudon DL, Ruddick ME, Wrenn B (2005) The Marketing Research Guide. Routledge, NY

Technorati (2008) State of the blogosphere report 2008 http://technorati.com/blogging/state-of-the-blogosphere/. Cited 8 Aug 2009

The JASON Project (2007-2009) http://www.jason.org/public/home.aspx. Cited 8 Aug 2009

The Nature Conservancy (2008) http://www.plantabillion.org/.Cited 14 Aug 2009

Tilden F (1957) Interpreting our heritage. University of North Carolina Press, Chapel Hill, NC

Zehr J, Gross M, Zimmerman R (1991) Creating environmental publications: a guide to writing and designing for interpreters and environmental educators. University of Wisconsin, Stevens Point Foundation Press, Inc, Wisconsin

Part III
Skills Building and Practical Applications

Chapter 10
Grouping Together Well

10.1 Introduction

A 'group' forms when two or more people interact in a way that allows each person to influence and be influenced by each other member of that group (Shaw 1981). Interaction differentiates a group from students sitting in a hall listening to a lecture, movie-goers viewing a film and passengers being carried by an elevator. These other assemblages lack the dynamics required to be considered a collection of people influencing each other, to be a real group.

The form of interaction within a group usually determines the types of motivation, group agenda(s) and interactive roles that exist therein. In this chapter, the broadest classification we'll use distinguishes formal from informal groups.

- **Formal** – Organized to do jobs with specific end goals. Formal groups are usually established within permanent organizations with enduring types of work and hierarchical command structures. Such groups often handle tasks and assignments within the overall structure of their parent organization.
- **Informal** – Formed by members who share a common goal or interest. Informal groups usually are more ephemeral than formal groups, often focused on resolving specific problems.

Environmental issues offer many interesting case studies of group formation and transformation of informal groups into organizations containing many formal groups. Many nonprofit groups were created to deal with specific issues of concern to the founding members. Later they expanded to include other related interests, which is to say they moved from informal to formal.

10.2 Why Do Groups Exist?

Groups form for reasons of shared need. A group's members wish to get something from the group that they are unable or unwilling to get for themselves. Attraction to groups falls into two categories.

- **Primary Motivation** It operates on an interpersonal level. People with primary motives join a group because they are attracted to qualities of other people within the group, have a vested interest in an outcome, or identify with the group's shared beliefs, attitudes and opinions. Members generally seek opportunities for more personal interaction with other members. Such social identification with a group serves to enhance the ego-function of the individual. There is a direct task orientation when there is primary motivation.
- **Secondary Motivation** It occurs through less strong feeling. This type of involvement serves as a means to an end, where the group meets a pragmatic, finite purpose. Joining a group for secondary motives allows one to network with desirable others, creating career-building or social circle opportunities. The prestige of membership – and not what the group could accomplish – is what is important. Colloquially, such membership can be referred to as 'padding the resume.'

10.3 Community Groups and Their Special Aspects

Environmental issues often create community groups. Citizens with primary motivations in seeing their environment protected organize for action. These groups tend to be short-lived and highly focused on the resolution of a problem. Their purpose is often singular and their tasks few. But, members can help themselves to be more successful if they pay attention to several qualities. A group's ability to overcome the negative qualities makes problem resolution more likely. Such group dynamics also shed light on the possibility of a more formal organization and permanent existence. While a communicator may be a central figure in organizing the group, it is imperative the group become self-sufficient. The following factors emphasize special points a communicator needs to be aware of when helping organize community groups.

10.3.1 Hegemony

Hegemony is the act of imposing a set of values on a situation or group. Though this concept is usually applied in the domain of international relations, it can also be fruitfully applied to group dynamics. Whenever people interact, each agent brings their own particular mindset and belief system to the situation. Forceful imposition of one's position is a form of dominance and implies the others' positions are of no consequence. Such coercion may only serve to alienate and close lines of communication. It is important, as a communicator, to realize that your values are yours. Do not assume others automatically share your values, your beliefs, your attitudes and your positions. Even if you are convinced that your views and values are the 'right' ones, it is up the community group to discover for themselves and therefore, incorporate those values and beliefs into their own cognitive framework.

10.3.2 Empowerment

A hegemonic view of group power says that empowerment, the ability to get a job done, comes from a higher authority. Some stronger group gives power to a lesser group. This view is most often counterproductive, because it is debasing and degrading. In contrast, a thoughtful communicator will assist a group in the discovery of their particular avenues of actions. Such a remedy will be more satisfying and more resistant to failure. Metaphorically, show them the map and let them find a path, with your guidance if necessary. Give them skills and help them find a way, through their own experience, to find a solution.

10.3.3 Revelation

Discovery is a process by which individuals in a group come to understand concepts or events from a new perspective. A conscientious, and ultimately successful, communicator helps people construct realizations for themselves. In the field of environmental interpretation, Tilden (1957) defines this essence of discovery as 'revelation.' Understanding is ideally revealed to a person or group such that they see connections for themselves without need for outsider explanations. While understanding the sense of what is being communicated is certainly important, it is the process of discovery that imparts deeper meaning to the message. When a group shares the experience of revelation, it is binding and builds cohesion.

10.3.4 Education as Intervention

All too often, education becomes a process of telling people what the educator thinks they should know. This ultimately can lead to indoctrination of dominant beliefs and give a distorted view of subordinate needs from a dominant perspective. It is critical to the true success of any educational effort to first be aware of what the target audience perceives its needs are before imposing education on them. Concepts from social marketing (see Chapter 19) and risk communication (see Chapter 18) can be important here. Educational interventions must square with the community's existing perspectives (Heimlich and Norland 1994).

10.3.5 Leadership and Dependence

Groups tend to form dependencies on leaders, and so need to learn ways to improve chances for self-reliance and independency. Leadership should, when possible, be grown from within a group. If a leader must come from outside, then this person may have to work hard at overcoming resentment and a feeling of intrusion. Cohesiveness of a group tends to be better maintained through an inside leader. But, if an outsider is able to become an accepted member of the group, their leaving can result in just-as powerful feelings of abandonment and potential for dissociation. Therefore, an outsider who gains leadership of a group must try not to become the central decision-maker of the group. Instead the outsider should lead as a guide and not as a ruler, especially if the group must continue to act long-term.

10.3.6 Openness

The degree to which interested persons are allowed to be involved in group process can be crucial to success of a group effort. Organizations that use public interaction must do so with the full intention of listening to what the audiences have to say. Audiences which are given a carrot on a stick of being involved, and then find theirs is just token involvement can quickly become hostile and non-cooperative. It is critical that you let the audience know at the outset what its role is in the interaction. Is it merely information-giving, in which they will have no power in the decision-making? Or will they be a full partner with the ability to say 'No.' The audience's role must clearly delineated for them and process decisions affecting group function need to be transparent.

10.4 Team Building Techniques

One of the givens for a professional communicator in today's world is working in 'teams.' This can be either within an organization, a loosely formed coalition, a task force or informal community. We've all heard horror stories about having to work in such groups, but the experience need not be negative. In this section, we offer some ideas on how to build a strong team, one that takes into account characteristics of its constituent individuals and actually uses them to forge stronger consensus and cohesion. After all, isn't that why we come together in groups?

Groups and organizations are part of how society has structured itself. People long ago made the decision to affiliate in a myriad of forms. Consider how many groups you belong to both within your social and professional lives. Consider the pros and cons of group membership:

- **Positives** Ideas and information generate better products
  Comfort and stress reduction
  Spreading of the workload
  Tasks can be achieved more quickly
  Different experiences and ways of thinking contribute to new ideas and solutions

- **Negatives** Hard to organize
  Hegemonic tendencies
  Uneven distribution of work and resentment
  Often no compromise situations develop
  Over-reliance on strong leaders

Groups are not simple entities to function within. But, issues and tasks they deal with are not simple either. Gestalt theory states that you get more from the whole group than you could from the sum of the individuals. While in many cases this is true, some team members may work better if left to themselves to work on a problem before inviting them to discuss it within the whole team. Knowing the preferences of the whole team will direct which method is best.

10.4.1 Group Climate

Groups function within an emotional and psychological environment. Dynamics of a group are heavily affected by this group climate. A climate conducive to profitable dynamics and to reaching satisfying outcomes can be helped by clear rules of participation, permissiveness and interactions. Who is allowed as members, what is allowed and not within group meetings, and expectations of how the group will work together are best stated upfront. Groups also produce a better climate for themselves to function in if they keep close tabs on member motivations, have agreement on their means of reaching agreement or consensus, and are vigilant in managing conflict.

Motivation can be central to a member's affiliation or quite tangential. Leaders and communicators working with groups should develop some sense of the motivational makeup of the group. Having this sense permits efforts to be more focused, by hanging on the needs and desires of group members. Likewise, having a known decision-making process in place make groups function more efficiently. Will formal rules be followed? Will votes be taken by secret ballot? Will decisions be made only after consensus is achieved? A member who knows the answers to these questions will be better able to take part.

Perhaps the most dangerous component of group dynamics is conflict. Conflict is inherent in any issue. It should be expected between and within all groups. But, conflict can be used positively to enhance group development and foster progress toward solutions. Not dealing openly with conflict inhibits group function, generates resentment and hostility, and may destroy relationships.

10.4.2 Building Relationships

Instead of succumbing to inevitable conflict, groups that flourish find their strengths and weaknesses and then make these obvious to all members. Groups are wise to make time early in their existences to identify shared characteristics and differences. This is particularly important because in most cases the group has come together because they feel they have a shared reality on either an issue or a task. They are oriented toward finding a solution. What comes out of early team-building exercises is the revelation that although they might share the same larger conceptual ideal, they have very different reasons for being there, and even different agendas for working on a task or issue.

Differences occur because of varying values, beliefs and opinions which are often masked behind symptoms, like too much focus on process and petty bickering. When we are faced with insufficient information to make a decision, we fill-in the gaps from our own perceptual background and experiences. In essence, our own view of reality dictates how we perceive the world and we assume erroneously that others see the world in the same way. We need to accept that we are hold different agendas. Acceptance fosters cohesiveness.

Here are some suggestions on group rules which have worked many times before:

- The best number of members for small-group interaction is five to seven. Informal groups with more than seven members usually become overly structured and more difficult to manage.
- Actively seek to discover strengths and weaknesses of the group as a whole and of members.
- Manage for conflict from the beginning. Accept that conflict is inevitable, but does not have to lead to problems. Identify sources of potential conflict and openly discuss them. Agree to disagree.
- Openly discuss the group's goals so that everyone knows what is happening. Allow each person to openly express their personal viewpoint. This will uncover major areas of difference. Do not be judgmental on another group member's viewpoint. You are trying to establish commonalties and differences to unify the group.
- Set-up a schedule. Define the group's time constraints and set milestones. Identify when group members can meet, and what time restrictions exist. Set realistic goals and objectives for the group.
- Get agreement on leadership and facilitation. Who will lead or direct? Set-up a known system for delegating tasks.
- Deal openly with issues of participation, permissiveness and cohesiveness. These can be sources of conflict. Decide how members of the group will participate. If members do not follow through with designated tasks, decide how they will be sanctioned. Openly discuss what is or is not allowed. Work toward group cohesion with everyone being part of the team, and not just individuals working together. Be on the lookout for groupthink (see Box 10.1).
- Faithfully follow decision-making criteria.
- Solicit and review alternative courses of action. Review the complete range of objectives and the values implicated by a decision.
- Closely examine the consequences of each alternative.
- Search for new relevant information.
- Incorporate all new relevant information.
- Reexamine the consequences of an inappropriate solution.
- Make detailed plans for implementation and action of the selected final decision.

> **Box 10.1** Groupthink
>
> Sometimes group processes grind to a halt or even spiral downward. When efficiency and progress deteriorate, 'groupthink' (Janis 1982) may have infected the dynamics between members. Groupthink is defined as 'a pathological mode of thinking within groups where concurrence seeking becomes so dominant in a cohesive in-group that it tends to override realistic appraisal of alternative forms of action' (Janis 1982). In a case of groupthink, members tend to become too selective in the information used to make decisions. Alternatives are ignored. Facts and opinions contrary to the group's stated position are ignored, while facts and opinions supporting the groups position are readily accepted without question.
>
> Symptoms of groupthink
>
> - Believing the group is invulnerable.
> - Rationalizing avoidance of warnings and threats.
> - Believing the group is completely moral and just.
> - Placing irrational pressure on member who expresses doubts.
> - Members censoring their own doubts.
> - Belief that members of the group are in unanimous agreement.
>
> Prevention of groupthink
>
> - Allow all members to be critical evaluators.
> - Discuss all viable options.
> - Assign one member to be the 'devil's advocate,' purposely expressing alternatives and criticisms of group positions.
> - Seek opinions of knowledgeable but uncommitted outsiders.
> - Separate into sub-groups to explore options.
> - Think rationally.
> - Assign one member to be a sounding board.
> - Consider the consequences of and processes used to make prior decisions.

10.5 Capacity-Building and Civic Agency

Effective organizations rely on strong, dedicated and skilled individual leaders in order to achieve their goals. The development of effective organizations and individual leaders is capacity-building. But, capacity-building goes beyond merely producing strong individuals. It also involves getting groups to work toward common ends. So, groups working toward environmental goals can be linked through capacity-building, as leaders network and environmental efforts become more comprehensive and collaborative. Powerful coalitions can be formed among groups if they know of each other's existence and recognize their common positions.

Coalitions are powerful entities often formed to resolve situations that otherwise might be irresolvable if acted on individually or by separate groups. Coalitions encompass multiple groups, using representatives from across identifiable viewpoints in a situation and ensuring every member of the coalition has equal status. The special factors used in understanding community groups all apply to coalitions. However, coalitions will most likely be formed of a larger and heterogeneous mixture of business/industry, agencies, and various community and special interest groups.

Environmental communicators can participate in capacity-building through these activities:

- Identify likely allies.
- Identify key people in key groups that share your agenda and role.
- Identify potential members and the reasons they may wish to join your group.
- Recognize potential barriers to forming coalitions. Develop strategies for dealing with inter-group conflicts.
- Develop strategies to make potential members aware of your group.
- Involve all coalition members equitably.
- Establish a shared vision.
- Set realistic goals and objectives.
- Involve all members in some way.
- Evaluate the levels of cooperation among coalition members.
- Make meetings meaningful.
- Share successes among all.
- When conflict comes, reveal it, do not conceal it.

In short, the forming of a coalition can lead to productive solutions. A communicator often acts as the facilitator or convener and helps keep the coalition focused and on point.

The term *civic agency* has a long history. As Boyte (2005) points out, it comes in various disguises such as citizen action, civic engagement, citizen agency, community organizing, and adult/community education. What it all comes down to is simple – working well with others to make something positive happen for the 'common good.' Mere public participation where governing bodies consult citizens to learn their views before making a decision does not rise to the level of civic agency, which is where citizens shape their own lives and communities. It is citizens that respect the capacities and resourcefulness of all their members in creative partnerships through negotiations, ownership of the issues, co-learning situations and self-empowerment. It is a bottom-up citizen movement as opposed to top-down authoritarian dictates. In a sense, it is a democracy of citizens working equally with the leadership to find solutions to civic questions and problems.

10.6 Managing Versus Leading

Are 'manager' and 'leader' synonymous terms? On the surface they are, but in essence, there is an exclusive difference. Management is transactional, in its pursuits of processes such as measurement, structures and procedures, and overall control of a system. Leadership, however, is transformational, intimately involving people skills to achieve buy-in, commitment, empowered creativity and encouragement for self-direction of individuals within the bounds of the team. All the while leaders work toward management goals, reinforcing dynamics of a group's structure and purpose. A good leader seeks out potential limits to organizational growth and then works to eliminate or minimize them before they hinder group progress. Group results are still a priority, but leaders help people do their jobs rather than managing the process. Chinese philosopher Lao Tzu is quoted as saying, 'When the sage's work is done the people will say, "We did it ourselves."'

Thus a rewards system is transformed into an intrinsically successful working of a system. The group, rather than its individuals, earn rewards and achieve results. An excellent manager lives with the notion that he or she may not always be readily recognized for the successful running of their system (Repenning and Sterman 2001). A good manager and leader therefore aims for working smarter not harder – increasing performance, effort and capability without working harder, and actually working with less stress and more individual fulfillment! Managers who have not mastered leadership skills are more apt to blame people for problems and not the system itself.

10.7 Some Management Theory Ideas

10.7.1 Contingency Theory

This approach can be summed up as 'it depends on the situation' management. If driving towards a singular goal in an urgent situation, an autocratic style of control management may be more preferable since time and resources may be a major constraint. If, however, a situation is longer-term management of a complex group, then participatory and facilitative leadership style is probably better employed.

10.7.2 Chaos Theory

We like to believe that we can control all our circumstances, yet observation shows us that chaos may be more the norm. Chaos theory predicts that whenever we think we can predict events and circumstances, then the unpredictable happens. Chaos, however, refers to dynamics of a system that apparently have no order, but in which there really is an underlying order. This is partly because of the complexity of the systems in which we live, both human built and natural. The more complex a system the more energy is needed to maintain it in a 'stable' form. Inevitably systems fail or more usually change to adapt to changing external pressures. Adaptation to change seems to be the only constant in our world. As such, successful managers and leaders need to understand the systems in which they work.

10.7.3 Systems Theory

In its fundamental nature, a system is a collection of all the parts of which it is made, and probably contains subsystems integrated together as a part of the whole. For example, an organization is made up of many administrative and management functions, services, groups and individuals. If one part of the system changes, then the overall system is often changes as well. We call this systemic, meaning relating to or affecting, the entire system. Systems theory and systems thinking are not the same as being systematic, which means being methodical. In the context of a communications campaign, systematic means setting goals and objectives, collecting and analyzing feedback, and then adjusting activities to achieve the goals more effectively. Systems theory is about how the system exists, thrives or flounders.

The most apparent systems we see everyday are biological. Our bodies are complex systems that interact in highly elegant and hierarchical ways to maintain homeostasis (called 'dynamic equilibrium' by biologists). We maintain our core body temperature, our blood sugar levels, the flow of oxygen during different activity levels, and myriad others systems on a continual basis. These systems are usually hierarchical with their own set of boundaries that interact with their connecting environments. As such, they maintain a high-functioning system that continually exchanges feedback among its various parts to ensure they remain closely aligned and focused on maintenance of the whole system. If any part of the system becomes weakened, the system adjusts to more effectively achieve its needs and compensate.

A mound of sand is not a system. If you remove a grain, or even a thousand grains, of sand, you are still left with a mound of sand. But, a mound of healthy soil contains a complex system of multiple minerals, dead organic matter, and many forms of organic life (bacteria, fungi, protists, worms, grubs and other insects that biodegrade organic matter). If you remove one of the main components then the soil quality and structure suffers, and ultimately leads to a plants decreased ability to thrive in its growth medium. Our organizational structures are similar. If a large park system had great leaders and staff for each of its departmental units but the individual departments do not talk well to each other, then the park's system would be prone to failure since problems would be isolated as part of each department rather than the integration of the whole system. The park system would not adjust and fail to reach dynamic equilibrium.

When we focus on a whole system, we can identify solutions that address as many problems as possible in the system. The positive effect of those solutions leverages improvement throughout the whole system, thus leading to the term 'systems thinking.' Small changes can create complex positive changes in the overall system. One of the major contributors to organizational systems thinking has been Senge (1990), who developed strategies to help organizations become successful through understanding the system in which they exist. Senge gave five competencies necessary to cultivate systems thinking in an organization or communication effort.

- **Systems Thinking** Uses the ideas presented above to analyze how organizations are structured and run.

- **Personal Mastery** This is where each individual in an organization continually clarifies and deepens their personal vision, of focusing energy, developing patience and of seeing reality objectively – nurturing critical perspectives.
- **Mental Models** This is where the individuals that make up the organization analyze the mental models (worldviews) that are deeply ingrained assumptions, generalizations or images that influence how we understand the world and how we take action.
- **Building Shared Vision** Vision is the one aspect of leadership which has inspired organizations for thousands of years. The whole group should contribute and buy into a shared picture of the future they wish to manifest as a group.
- **Team Learning** Senge asks 'How can a team of committed managers with individual IQs above 120 have a collective IQ of 63?' Learning has to occur for the team as a unit, not just within the individuals involved.

10.7.4 Self-Directed Teams

Add to systems thinking the notion that a team is best directed by members of the team and you get the concept of 'self-directed team.' In self-directed teams, members share interrelated tasks, bear collective responsibility for products and goals, determine how to apply skills within the group to accomplish tasks, and are evaluated in terms of the team rather than individuals (Bishop and Scott 2000). Self-directed teams have a lot of latitude in determining their own distribution and pace of work. Sharing of management functions builds commitment and fosters leadership development of all team members.

10.8 Emotional Intelligence

Effective leaders have been shown, by and large, to possess a high degree of what has come to be known as *emotional intelligence*. Emotional intelligence is the ability to manage one's own reactions and one's relationships with others. It is described by five skills:

1. **Self-Awareness** The ability to recognize and understand your moods, emotions and drives, as well as their effect on others.
2. **Self-Regulation** The ability to control or redirect disruptive impulses and moods.
3. **Motivation** A passion to work for reasons that go beyond money or status.
4. **Social Skill (Social Intelligence)** means proficiency in managing relationships and building networks; an ability to find common ground and build rapport. Goleman (2006) emphasizes that humans have a built-in bias toward empathy, cooperation and altruism – provided we develop the social intelligence

to nurture these capacities in ourselves and others. This quality has been equated to Gardner's *interpersonal intelligence* (Gardener, 2006). People exhibiting a high degree of interpersonal intelligence are skilled at assessing the emotions, motivations, desires and intentions of those around them. Consequently, they are good at communicating verbally and non-verbally, able to see different perspectives easily, create positive relationships, and are good at resolving conflict in groups. From a clinical perspective, the brain structures of this intelligence have been well described. Lobotomy patients and those with damage to the frontal lobe have shown negative personality changes and a reduction in their ability to interact well with others, even if they were able to do so well before. Goleman (2006) states, 'Our reactions to others, and theirs to us, have a far-reaching biological impact, sending out cascades of hormones that regulate everything from our hearts to our immune systems, making good relationships act like vitamins – and bad relationships like poisons.... [Social intelligence] surprisingly [explains the] accuracy of first impressions, the basis of charisma and emotional power, the complexity of sexual attraction, and how we detect lies.'

5. **Empathy** The ability to understand the emotional makeup of other people, skill in treating people according to their emotional reactions. *Empathy* is the capacity to recognize or understand another's state of mind or emotion, the ability to 'put oneself into another's shoes,' a skill greatly diminished in modern society. A major component of empathy is *compassion,* which is a profound human emotion that gives rise to an active desire to alleviate another's suffering. Davidson states, 'Many contemplative traditions speak of loving-kindness as the wish for happiness for others and of compassion as the wish to relieve others' suffering' (Lutz et al. 2008).

It's important to emphasize that building one's emotional and social intelligence cannot, will not, happen without a sincere desire and concerted effort. Nothing great can be achieved without enthusiasm, so if one wants to really become empathic, then high emotional intelligence must be developed.

10.8.1 Trust as the Cornerstone of Empathy

Part of the empathy process is establishing trust and rapport, which helps in having sensible and mindful discussions. Establishing trust is about listening and understanding without judging – not necessarily agreeing. A useful focus when listening is to try to understand the other person feelings, and to discover what they want to achieve. Of all the communications skills, listening is arguably the one which makes the biggest difference. Listening does not come naturally to most people, so we need to work hard at it; to stop ourselves 'jumping in' and giving our opinions.

Active listening includes doing something that demonstrates you are listening and have understood, such as:

- Giving non-verbal cues to demonstrate you are paying attention – nodding, making eye contact, making facial expressions appropriate to what is being said.
- Reflecting back the main points and summarizing what has been said.

Being aware of and learning how to develop social and emotional intelligences brings numerous benefits to individuals and groups. As communicators helping to promote social and emotional learning, we help develop social-emotional intelligence in our audiences. This in turn has synergistic effects on other socio-cultural characteristics to promote pro-social bonding. High-quality environmental communications can help foster a more knowledgeable, responsible, caring, productive, ethical and contributing citizenry. Beneficial person-centered social and emotional learning competencies embrace (Zins et al. 2004):

- **Self-Awareness** Identifying and recognizing emotions, accurate self-perception, recognizing strengths, needs and values, self-efficacy and spirituality.
- **Social Awareness** Perspective taking, empathy, appreciating diversity and respect for others.
- **Responsible Decision-Making** Problem identification and situation analysis; problem-solving; evaluation and reflection; and personal, moral and ethical responsibility.
- **Self-Management** impulse control and stress management, self-motivation and discipline, and goal-setting and organizational skills.
- **Relationship Management** communication, social engagement and building relationships; working cooperatively, negotiation and conflict management; and help seeking and providing.

10.9 Formats for Presenting Information to Groups

As an environmental communicator, many of your messages will be crafted for and presented to groups. Recall audience analysis purposely groups people. When your audience has already gathered itself together under a philosophical umbrella and given itself a label, as organizations have done, your process of message planning and delivery should be assisted. Also, as you are preparing to communicate with a group, it behooves you to find out what form of interaction your presentation can take. The communicator can take advantage of the kind of group by varying the format to meet the end goal. Here are some examples of form and function of group presentation:

10.9.1 A Speech, Film or Demonstration

This gives information in an organized way but does not allow an audience a chance for open discussion. Such a presentation is most successful when delivered by a person who knows the subject thoroughly and can present it engagingly and with visual impact. Combine questions and group discussion to get a modicum of participation. On the positive side, this format allows a lot of information to be disseminated quickly.

10.9.2 Brain-Storming

This open-ended, non-judgmental technique gets many ideas out quickly. Members of the audience throw out an idea on the topic under discussion while one person records them. It is important to accept any idea, no matter how wild. A discussion of the ideas and how best to organize them then allows the group to see which ideas are worth pursuing further. Brain-storming can help produce a cohesive group and takes advantage of the 'safety' of a group setting.

10.9.3 Buzz Sub-Groups or Small Discuss Sub-groups

To use these sub-groups well, divide a large group into sub-groups of 4–10, preferably around tables. Designate a facilitator and recorder for each group. Make the topic discussion clear and focused. After allowing 5–15 min of discussion, bring the whole group together and get a summary report from each group. End with general discussion of all the points raised. This way everybody has a chance to take part in the discussion. Buzz sub-groups are good for getting commitment to action, but hinders in-depth discussion.

10.9.4 Role Playing

To use role playing, give each participating member a card briefly explaining the character they will represent. Indicate clearly the points of view to be taken though not the exact words they are to say. Try to choose players who will take criticism in good spirit. Take the players aside and tell them to have fun with this exercise, but to remain in character during the discussion. Then, let them launch the role-playing session. Stop the role playing before interest wanes and immediately follow-up with an open discussion of what just occurred. The main purpose here is to get people to adopt other ways of thinking. This technique helps in understanding attitudes and opinions of others. Role playing does not, however, always provide new information.

10.9.5 Panel Discussion

A moderator guiding three to five panel members with different views is a good technique for looking at various perspectives on a topic, in front of an audience self-selected to be interested in the topic. Keep talk between panel members free-flowing and conversational. The moderator should not be a central figure in the discussion, but should only invite comments and reiterate or ask questions if the discussion starts to become dry. Allowing a free flow of questions from the audience makes the discussion more inclusive and interesting.

10.9.6 Colloquy, or Talk-Show Format

This technique is similar to a panel discussion, with a bit more influence by the guide and more discussion from members of the audience encouraged. This is often used in free information exchange conferences, in early stages of coalition building, or issue resolution to identify community concerns.

10.9.7 Symposium

In a symposium, each person on a panel of experts delivers a speech/presentation to the audience. A symposium chair introduces the speakers, provides transitions, and may provide occasional summaries. Usually symposia are unidirectional communication modes. An audience needs to understand that its role is mostly limited to receiving information.

10.10 Conclusion

The concept of a group is important to the environmental communicator, because audiences most often consist of individuals who influence one another. Formal organizations, temporary project teams, and ad hoc community groups are just some of the groups that are common in resource management issues. Regardless of whether the communicator is a member of the group or communicating from outside the group, understanding the internal dynamics of the group is imperative for understanding and predicting how members will transfer and understand messages.

10.11 Case Study: Environmental Group Formation. Taiwan's Environmental and Sustainability Non-governmental Organizations

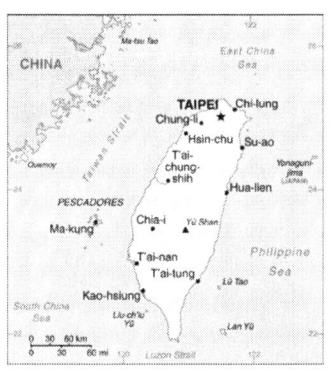

Taiwan broke away from mainland China's rule in 1949. The island was governed under martial law until 1987, when democracy began to sprout. Despite limited international recognition, Taiwan has developed into an economic powerhouse, with 2008's top two global manufacturers of netbook-class computers both headquartered in Taipei. Democratization precipitated an 'explosion' of non-governmental organizations (NGOs) working on social issues, from human rights, culture, international relations, and the environment (Lin et al. 2005). By 2000, Taiwan was home to 20,000 NGOs and began a fresh era of global networking.

On environment and sustainability issues, the Taiwanese NGO community is particularly robust, with leading groups such as Society of Wilderness, Green Formosa Front, Homemakers' Union and Foundation, Taiwan Environmental Action Network, and Earth Passengers (Tsai 2008). Founded in 1996, Society of Wilderness is Taiwan's largest NGO, with about 10,000 members. Though the organization relies on volunteers, their corps displays exemplary place-based knowledge and interpretive skills (Morgan et al. 2006). In 2002, 20 NGOs formed a coalition to send representatives to the UN World Summit on Sustainable Development in Johannesburg, South Africa. The move was so revelatory and empowering that the Cabinet-level Environmental Protection Agency added their support (Tsai 2008).

Still, by 2005, indicators showed only modest depth and breadth in civic participation through NGOs. Almost two-thirds of Taiwanese citizens displayed an external locus of control, while only 28% had joined any NGO and less than 15% had volunteered with any NGO (Lin et al. 2005). But, civic participation continues to expand; by 2008, another 10,000 NGOs had been established (Tsai 2008).

Credit: Taiwan map. The World Factbook.

References and Further Reading

Adler NJ, Gunderson A (2007) International dimensions of organizational behavior, 5th edn. South-Western College Publishers, Cincinnati, OH
Beckhard R, Pritchard W (1992) Changing the essence: the art of creating and leading fundamental change in organizations. Jossey-Bass, San Francisco
Bishop JW, Scott KD (2000) An examination of organizational and team commitment in a self-directed team environment. J Appl Psych 85(3):439–450
Boyte HC (2005) Everyday politics: reconnecting citizens and public life. University of Pennsylvania Press, Philadelphia, PA
Boyte HC, Shelby D (2008) The citizen solution: how you can make a difference. Historical Society Press, Minnesota
Connolly P, Lukas CA (2002) Strengthening nonprofit performance: a funders guide to capacity building. Fieldstone Alliance, St. Paul, MN
Conrad C, Poole MS (2004) Strategic organizational communication into the twenty-first century, 6th edn. Wadsworth, Belmont, CA
Daniels TD, Spiker BM, Papa MJ (1997) Perspectives in organizational communication, 4th edn. Brown & Benchmark, Madison
Daft RL (2009) Organization theory and design, 10th edn. South-Western College Publishers, Cincinnati, OH
Fettig L (2007) The ABCs of development: it's about building capacity. AuthorHouse, Herausgeber
Forsyth DR (1998) Group dynamics, 3rd edn. Brooks/Cole, Pacific Grove, CA
Gardener H (2006) Multiple intelligences: new horizons in theory and practice. Basic Books, New York
Gastil J, Levine P (2005) The deliberative democracy handbook: strategies for effective civic engagement in the twenty-first century. Jossey-Bass, San Francisco

References and Further Reading

George JM, Jones GR (2007) Understanding and managing organizational behavior, 5th edn. Prentice-Hall, London

Goleman D (1995) Emotional intelligence. Bantam Books, New York

Goleman D (2006) Social intelligence: the new science of human relationships. Random house, New York

Heimlich JE, NorlandE E (1994) Developing teaching style in adult education. Jossey-Bass, San Francisco

Janis IL (1982) Groupthink: psychological studies of policy decisions and fiascoes, 2nd edn. Houghton Mifflin, Boston

Kennedy D, Barker JA (2008) Putting our differences to work: the fastest way to innovation, leadership, and high performance. Berrett-Koehler Publishers, San Francisco

Kolb DA, Osland JS, Rubin IM (1995) The organizational behavior reader, 6th edn. Prentice Hall, Englewood Cliffs, NJ

Lutz A, Brefczynski-Lewis J, Johnstone T, Davidson RJ (2008) Regulation of the neural circuitry of emotion by compassion meditation: effects of meditative expertise. PLoS ONE 3(3):e1897. doi:10.1371/journal.pone.0001897

Lin T, Liao J, Fields A (2005) An assessment of civil society in Taiwan, transforming state-society relations: the challenge, dilemma and prospect of civil society in Taiwan. National Sun Yat-sen University & World Alliance for Citizen Participation, Kaohsiung

Moorehead G, Griffin RW (1997) Organizational behavior: managing people and organizations, 5th edn. Houghton Mifflin, Boston

Morgan M, Lin HS, Chou J, Wu H (2006) An interpretation specialization continuum of environmental volunteerism in Taiwan. J Interpret Res 11(2):7–20

Oyster C (1999) Group dynamics. McGraw-Hill, Boston

Pearse M, Smith J (1990) Community groups handbook, 2nd edn. Journeyman/Community Development Foundation Publications, Nottingham

Repenning NP, Sterman JD (2001) Nobody ever gets credit for fixing problems that never happened. Calif Mgmt Rev 43(4):64–88

Robbins SP, Judge TA (2008) Organizational behavior, 13th edn. Prentice Hall, Upper Saddle River, NJ

Schermerhorn JR Jr, Hunt JG, Osborn RN (2008) Basic organizational behavior, 10th edn. Wiley, New York

Senge P (1990) The fifth discipline. Doubleday, New York

Shaw M (1981) Group dynamics: The psychology of small group behavior, 3rd ed. McGraw-Hill, New York

Skinner S (2006) Strengthening communities: a guide to capacity building for communities and the public sector. Community Development Foundation Publications, London

Soska T, Johnson Butterfield AK (2005) University-community partnerships: universities in civic engagement. Routledge, New York

Stewart GL, Manz CC, Sims HP (1998) Team work and group dynamics. Wiley, New York

Stewart GL, Brown KG (2008) Human resource management: linking strategy to practice. Wiley, New York

Tilden F (1957) Interpreting our heritage. University of North Carolina Press, Chapel Hill, NC

Thames B, Webster DW (2009) Chasing change: building organizational capacity in a turbulent environment. Wiley, Hoboken, NJ

Tsai J (2008) Global networking NGOs promote sustainability. Taiwan Times, Government Information Office, Republic of China, Taipei

Twelvetrees A (2008) Community work, 4th edn. Palgrave Macmillan, Basingstoke

Watson D (2007) Managing civic and community engagement. Open University Press, Maidenhead

Zins JE, Weissberg RP, Wang MC, Walberg HJ (eds) (2004) Building academic success on social and emotional learning: what does the research say? Teachers College Press, New York

Chapter 11
Differing Ways of Thinking and Doing

11.1 Introduction

For any form of communication to impart a message that triggers learning within its recipient, a communicator must take into account the diverse ways people learn and make use of media. These different styles of personality, learning and coping often may conflict when people work together. Personality styles affect people's interactions because of different ways information is processed. Similarly, learning styles affect how people prefer to receive information. Richard Felder (2010) comments:

> '[People] have different learning styles characteristic strengths and preferences in the ways they take in and process information. Some tend to focus on facts, data and algorithms; others are more comfortable with theories and mathematical models. Some respond strongly to visual forms of information, like pictures, diagrams and schematics; others get more from verbal forms written and spoken explanations. Some prefer to learn actively and interactively; others function more introspectively and individually.'

11.2 Personality Types

Learning is affected by personality and personality is heavily influenced by learning over one's lifetime. This interplay produces differences in how each individual prefers to process information and to deal with others. It is when we work in groups that personality types can conflict. Understanding different personality types should help everyone become tolerant of each others differences.

Our intent here is to do no more than introduce these ways of categorizing personalities and lead you to instruments that will help develop your understanding of different personalities. Communicators may garner an understanding of how people interact, deal with problems, determine importance, and look at the world. People are rarely static in how they react to the world. Reactions depend on the situation. Still, most people have a preferred way, often unconsciously so, of reacting to situations and other people. Understanding why people react the way they do may reduce conflict and improve message efficacy.

11.2.1 Satir Modes

How do you view life? Satir modes (Satir 1964; Elgin 1980) describe how individuals communicate and make use of presupposition statements. Can you identify your mode? Notice Satir modes can easily conflict with each other. Recognizing the different modes that may exist in a group will help in warding off unexpected conflict based solely on personality traits.

- **Blamer** – Such a person often feels alienated and is convinced that most people are oblivious to their needs and feelings. The blamer usually reacts with demonstrative dominance. They usually criticize others and externalize problems to other people. A blamer frequently uses statements such as 'why do you do that,' 'you always do that,' and 'why spoil everything.' Two blamers communicating tend to spread conflict readily.
- **Placater** – The person who typifies this mode is afraid of upsetting people and can go to great lengths to avoid conflict and alienating people. They are usually distant, anxious and frequently use statements like 'oh, you know me, I don't care' and 'whatever you do is fine by me.' Often placaters do care, and the situation isn't fine by them and they get worked up because no one has realized it. Two placaters communicating tend not to reach easy quasi-consensus since neither will make a decision for fear of offending the other and hence losing support.
- **Leveler** – This person's *modus operandi* is to be open and tell it as it is! Sometimes, this occurs bluntly. Usually their verbal and non-verbal language match. Two levelers tend to communicate well since each is being honest. But, if a person has adopted a leveler mode and is not honest, a conflict escalation may result because the true leveler is confused and becomes defensive because of the 'phony' leveler.
- **Computer** – This person operates in an emotionless state. This mode also keeps non-verbals are at a minimum (like Mr. Spock or Data from the Star Trek series and films). If in doubt what to do or say, this mode offers the best option. There is less likelihood to misunderstand this mode or even a need to become defensive against it. Computer mode is less likely to lead to conflict situations. It will probably help to reduce conflict, being ultimately logical and 'to the point.' Still, a lack of outgoing cues can make the recipient feel a little alienated. A person with preference for computer mode tends to fear expressing feelings and emotions. Statements from this type of person would include 'one would think that...' or 'it can be safely assumed that...' or 'obviously, no cause for alarm.'
- **Distracter** – This state promotes panic. A person in distractor mode cycles through all the other four states. Non-verbals are also somewhat confusing since they often do not match the words being spoken and frequently change.

11.2.2 Myers–Briggs Personality Typing

Myers–Briggs Type (Introduction to Type 1976) is perhaps the most widely used personality instrument. First designed by Carl Jung, its scales are four continuums of action and thinking, and most of us will lay along them rather than be as polarized as the descriptions. Like the Satir modes, recognizing different personality profiles helps in diffusing conflict before it occurs.

- **Extrovert to Introvert (E to I)** – Contrary to the common meanings of these terms, this factor is not about being shy or outgoing. Rather it is about how we focus our attention and draw energy from interpersonal contact. An extroverted person will usually focus their attention outward to people, thus engaging people in his or her thought processing. An introvert, however, tends to remain quiet and focus any thinking internally until they are ready to speak. Conflict arises when each type feels the other is negating their thinking. Introverts mull problems over in their head until they feel they have a definitive answer and then will share conclusions aloud. Extroverts, however, will probably speak straight away and start running through options to resolve a problem. The extrovert is merely processing (thinking) out loud and not really negating what an introvert has said. Yet, an introvert is more than likely to perceive the extroverted person is not listening or being negative. They may both arrive at the same decision, but the process preferences are different. Knowing this helps resolve the seemingly silly conflicts that often arise from feeling that someone is ignoring your opinion.
- **Sensor to iNtuitive (S to N)** – This dimension aligns with how we acquire and use information. Sensors use direct factual information and their five senses to gather information for decision-making. Intuitives rely on creative imagination, inspiration and intuition to arrive at a decision. Avoiding conflict here calls for sensors to trust and rely on intuitives, and for intuitives to accept that sensors need sound reasoning to accept their arguments.
- **Thinker to Feeler (T to F)** – This category deals with the rationale we use to make decisions. The thinker will use objective and logical information to make a decision – a sort of cost-benefit analysis. The feeler will use personal values to make a decision. This might imply that feelers can be dogmatic in their approach, refusing to budge on an issue that they feel goes against their values. Thinkers, however, might be seen as more willing to compromise in order to gain a consensual decision for the benefit of others, even though it may go against their personal values.
- **Judger to Perceiver (J to P)** – Are you planned and organized or spontaneous and flexible? People who are judging tend to need to know where every minute is going and how it is going to be spent before they do something. Perceivers tend to go wherever winds of change take them. If you have ever been on vacation with a polarized judger, you will know just where you are going to stay and what you will probably eat, 2 weeks before you actually get there. The same vacation with a full perceiver will be an minimally planned adventure, with you not knowing where you will be 1 h hence.

Though knowing your own preferences can help you understand yourself, deeper value for communicators rests in understanding those with opposite preferences. You can understand why you may be having difficulty with message reception, and then be able to work through any conflicts.

11.2.3 Enneagrams of Personality

Hudson and Riso (2003) discuss enneagrams as a system of explaining nine personality types and their interactions. An enneagram is a nine-sided geometric figure. The name is borrowed for this typology to diagram nine personality types and the ways they can intersect. Along each of the nine lines, each representing an 'enneatype,' is a continuum between an intergrative (healthy) display or a disintegrative (destructive) display.

Enneatypes represent basic patterns, and individuals are unique variations among those patterns. For insight with this model, you should not just pick out the profile you would like to be! If, for example, without testing, you thought you were a Type 3 because you see yourself as self-assured and ambitious, be aware that you also need to identify that you can also be narcissistic and psychopathic within that type. Each person has a basic type, and then can have two support wing types. The overall pattern dictates how a person is likely to view the world and react to situations that occur. Understanding both the positive and negative traits of each type, and especially how they can interact with your own, helps to reduce conflict in interpersonal relations (Pearce and Brees 2007).

The nine enneatypes are:

1. **Reformer** Idealisitic and orderly (integrative) to perfectionistic and intolerant (disintegrative)
2. **Helper** Concerned and helpful (integrative) to possessive and manipulative (disintegrative)
3. **Motivator** Self-assured and ambitious (integrative) to narcissistic and psychopathic (disintegrative)
4. **Artist** Creative and individualistic (integrative) to introverted and depressive (disintegrative)
5. **Thinker** Perceptive and analytic (integrative) to delusional and paranoid (disintegrative)
6. **Loyalist** Likable and dependent (integrative) to dogmatic and masochistic (disintegrative)
7. **Generalist** Accomplished and extroverted (integrative) to excessive and manic (disintegrative)
8. **Leader** Powerful and expansive (integrative) to dictatorial and destructive (disintegrative)
9. **Peacemaker** Peaceful and reassuring (integrative) to passive and repressed (disintegrative)

As in Myers–Briggs Typing, it is essential to understand yourself as well as others, and how different personality traits and preferences can conflict. Recognizing potential for conflicts before they fester makes for much easier conflict management. It is also necessary to be honest with yourself when taking any personality test. We would love to all be perfect, but understanding your weaknesses makes you strong and resilient.

11.3 Learning and Coping Preferences

In the academic literature on education, numerous different learning styles and even different intelligences can be found. We overview three of these concepts to indicate how pervasive they are and to begin to elucidate how a communicator needs to be aware of these different preferred ways of thinking, especially when doing personal presentations or face-to-face workshops. Awareness of differing ways of thinking and doing can help a communicator reduce the noise associated with encoding and decoding of messages. Beside individual differences, one also needs to practice multiculturalism, since different ethnic groups have embedded ways of thinking and learning.

11.3.1 Field Dependent vs. Field Independent

Some people can easily find a graphical line hidden in a maze of other lines. Others, no matter how hard they try, cannot even begin to find a hidden figure. It has nothing to do with intellect, but rather how we perceive the world. Those that have a hard time finding particular data buried in a maze of similar data can be classified as 'Field Dependent,' or global thinkers, while those that seem to easily focus on a complex field of information and find discrete parts are 'Field Independent,' or analytical thinkers. Of course, those that fall in the middle will display traits from both ends of the continuum. While most of us will in the middle of this measure, we will discuss the extremes and how these types of people differ in their work and social thinking. Again, neither one is better or worse than the other – they are just different preferences in how they manage information and socialize.

As you read through the table above, notice how different the two ends of the continuum are. Most people tend to use the mode with which they feel most comfortable. If other people are not comfortable with that mode, then strained interactions may result, or the audience will not be attentive (Table 11.1).

11.3.2 Gardner's Multiple Intelligences

Gardener (1993, 2006) has developed a widely respected theory that we all have at least eight unique intelligences. His body of work holds we all have these intelligences, but

Table 11.1 Characteristics of field dependent and field independent thinkers

	Field dependent	Field independent
Relationship to peers	Likes to work with others to achieve a common goal Likes to assist others Is sensitive to feelings and opinions of others	Prefers to work independently Likes to compete and gain individual recognition Task oriented; is inattentive to social environment when working
Personal relationship to instructor	Openly expresses positive feelings for instructor Asks questions about instructor's tastes and personal experiences Seeks to become like instructor	Rarely seeks physical contact with instructor Formal forms of address Interactions with instructor are restricted to tasks at hand
Instructional relations to instructor	Seeks guidance and demonstration from instructor Seeks rewards which strengthen relationship with instructor Is highly motivated when working individually with instructor	Likes to try new tasks without instructor's help Impatient to begin tasks; likes to finish first Seeks nonsocial rewards
Characteristics of learning program	Objectives and global aspects of program are carefully aligned Concepts are presented in humanized or story format Concepts are related to personal interests and experiences of learners	Details of concepts are emphasized; parts have meaning of their own Deals with math and science concepts Based on discovery approach
Content	1. Social abstractions: Field-dependent program is humanized through use of narration, humor, drama, and fantasy. Characterized by social words and human characteristics. Focuses on lives of persons who occupy central roles in the topic of study, such as history or scientific discovery. 2. Personalized: Ethnic background of learners, as well as their homes and neighborhoods, is reflected. The instructor is given the opportunity to express personal experiences and interests.	1. Math and science abstractions: Field independent program uses many graphs and formulas 2. Impersonal: Field-independent program focuses on events, places and facts in social studies rather than personal histories.
Structure	1. Global: Emphasis is on description of wholes and generalities; the overall view or general topic is presented first. The purpose of use of the concept or skill is clearly stated using practical examples. 2. Rules explicit: Rules and principles are salient. (Learners who prefer to learn in the field-dependent mode are more comfortable given rules than when asked to discover the underlying principles for themselves.) 3. Requires cooperation with others: The program is structured in such a way that learners work cooperatively with peers or with the instructor in a variety of activities.	1. Focus on details: Details of a concept are explored, followed by the global concept. 2. Discovery: Rules and principles are discovered from study of details; the general is discovered from the understanding of the particulars. 3. Requires independent activity: The program requires learners to work individually, minimizing interaction with others.

Source: Biocognitive processes in multicultural education (Tables 1–3, pp. 205–207) by Casteneda and Gray (1974). Learn more about ASCD at www.ascd.org.

11.3 Learning and Coping Preferences

some are more expressed than others. Gifted people may have a higher than average score on a specific intelligence, yet score lower on others. Overall, most people score around average on a few of the intelligences with one or two higher and one or two lower. Gardner's multiple intelligences are:

- **Spatial Intelligence** People who are high in this ability tend to think in pictures and create vivid mental images to retain information. They enjoy dealing with visual stimuli and will probably work in positions that emphasize their visualization skills.
- **Linguistic Intelligence** People who are good at languages have high ability in this category. They can listen and speak well and tend to think using words. They probably work well in positions where speaking is a primary part of the work.
- **Mathematical Intelligence** These people think in logical and numerical patterns, and are good at synthesizing information. They are also research oriented and like to unravel problems.
- **Kinesthetic Intelligence** Intelligence of movement means people high in this category will prefer to express themselves through interaction with the space around them. They make good dancers and athletes, and are adept at manipulation of objects. Technicians who use sophisticated equipment will also be high in this category.
- **Musical/Rhythmic Intelligence** People who are adept at music or can pick up a tune the first time they hear it are high in this category. They tend to be highly tuned to sounds around them and react accordingly. They tend to think in musical patterns.
- **Interpersonal Intelligence** High empathy and an ability to easily view other perspectives emphasize this category. They work to manage conflict. They are also highly intuitive and perceptive in nonverbal use. Counselors, motivational speakers, and clergy will probably have this as a primary intelligence. According to Goleman (2006), this is perhaps the most crucial intelligence for an effective communicator.
- **Intrapersonal Intelligence** A person with a higher level of this intelligence will be highly self-reflective and be cognizant of the dynamics of relationships. They are most evaluative and aware of their own inner workings.
- **Naturalistic Intelligence** This is becoming a much discussed intelligence ever since Richard Louv (2005) published *Last Child in the Woods: Saving our Children from Nature-deficit Disorder* and is possibly the most controversial of Gardener's intelligence. People high in this intelligence have a high level of sensory perception and are adept at recognizing patterns and relating them to natural systems, and as such are highly aware of their surroundings. They also tend to be more in tune with nature and natural settings. Naturalists are good at categorizing and cataloguing information.

Persons with a higher emphasis in a particular intelligence will prefer to use that mode in their communication. People play to their strengths. Being aware of your own modes and being sensitive to other modes will help when interacting to build groups and process information as a group.

11.3.3 Learning Styles

While multiple intelligences emphasize what modes we work best in, the following learning styles emphasize the kind of ways we prefer to learn. Google 'learning styles' and several useful hands-on tests can be found to identify your specific preference of learning. Many more learning style instruments and references to others can be found in Bennett (2006). Learning styles include:

- **Visual** These learners prefer to see what is happening. Visual aids are most desired, and they need to have a full view of a speaker to garner nonverbal nuances. They will also take detailed notes to visualize and absorb the information.
- **Aural** Aural learners prefer to hear information. They are tuned into a speaker's voice patterns and inflections. Audio-books are an ideal medium for these learners.
- **Print** Learners with this preference prefer to read their information from text and usually do so alone.
- **Kinesthetic** These people are usually fidgety and need to move around. Sitting still bores and distracts them. When speaking, it is best to allow them to pace slowly or manipulate objects. As a learner, they need to have something physical to do. They are animated and love to explore.
- **Haptic** People in this category prefer to feel objects and work hands-on. They learn through doing and touching. This group is similar to kinesthetics, except they have more patience and pay attention to specific objects for longer periods.
- **Olfactory** This category prefers odors and smells to recall information and trigger memory. They are statistically a smaller group than the others, but possess a powerful way of remembering information.

Learning styles indicate different preferred modes of interaction. While some will be willing to sit and listen to a speaker, others need to be doing something active in order to be attentive. Assessing your audience and varying your presentation techniques will result in a more attentive audience.

11.4 Accommodating People with Disabilities

Audience members with disabilities may require accommodations. More than that, people with disabilities need to be considered within the makeup of your audiences and included just as any other group is. The United Nations defines disability as 'long-term physical, mental, intellectual or sensory impairments that, in the face of various negative attitudes or physical obstacles, may prevent those persons from participating fully in society' (Byrnes et al. 2007). Of the 650 million persons with disabilities, most are poor and most live in developing countries. They constitute the world's largest minority. One of the basic rights they are often denied is access to information.

Discrimination against people with disabilities is rarely explicit. Rather, social conventions and structures subtly conceal people with disabilities, preventing them from obtaining information, education and employment in equitable ways. Inclusive communication campaigns can encourage acceptance of disability as just another aspect of diversity, alongside gender, religion, ethnicity, and so forth. For now, people with disabilities are more like to live in poverty; people who are poor are more likely to be disabled. And, the poor and disabled are more likely to live in unhealthy environments. Environmental communicators' can be sensitive to such marginalizing.

The World Bank offers practical tips for including 'visibility of disability' in communication products (Werneck 2005):

- Use proper communication formats and channels (such as large print, Braille, sign language, plain language, closed captioning video, audio-described videos) to establish effective two-way communication with people with disabilities.
- Choose physically accessible locations as sites for meetings.
- Use proper language and culturally acceptable terminology (ask local disabled people organizations what is locally appropriate).
- Include children and adults with disabilities in curricula, textbooks and storybooks for early childhood, primary, secondary and adult education. Develop journalism and broadcast capacity for, about and with people with disabilities. Encourage government agencies, non-governmental organizations and private sector advertisers to include children and adults with disabilities in their commercials and public service announcements. Support positive inclusion and portrayals of people with disabilities in dramas, documentaries and commercial films.

The modern and global ubiquity of information/communication technologies serves up ample opportunities for inclusion, for making information more accessible to disabled people. Chisholm and May (2009) apply concepts from architecture and product design to Wed design, arguing that as the Internet develops the profiles of user groups is becoming more and more diverse. Incorporating design elements originally developed for disabled persons is now good business, they say, because those practices are necessary to keep up with growth in audience groups such as older adults (aged 50+), mobile device users, those using audio-only, young children and those who do not even speak the written language of a particular site.

11.5 Conclusion

A lot of research has gone into understanding how people think and learn. Ideas from personality typing, multiple intelligences, and various learning styles can offer insight to communicators asking critical questions such as:

- Why are my meetings sometimes fraught with tension and misunderstandings? And, yet at other times they flow well, even when I use the same techniques. What's up with that?
- Why are some participants so frustrated, even hostile?
- Why doesn't the audience understand the simple schematic diagrams I've given them?
- Why does the audience want a visual when I explained it clearly already?
- Why is the audience so unresponsive even though the lecture was well-planned?
- Why don't those people respond right away, instead of putting their heads down and reading everything?

Much of this chapter's concepts are more applicable to face-to-face activities, such as personal presentations and community-based educational interventions. They are more limited when using mass media. The core question: What factors may come into play which would stop this audience from comprehending my message?

11.6 Case Study: Multiple Intelligences and Learning Styles

Rosenberg (2003) asked if knowing the multiple intelligence profile (MIP) of students could predict student success, and help to identify student motivation, within a lecture-based college biology class. He found student success was not closely tied to MIP, due to the prior development of varied and often ineffective study skills. The case suggests that mismatched study skills and learning preference can linger, when students are originally taught to study in ways which do not meet their learning styles. Motivation, on the other hand, was tied to the teaching style used and MIP for each student. While alignment of study skills to MIP needs to be studied more, this study points to benefits from helping people become motivated.

Meanwhile, other studies (e.g., Terregossa et al. 2009) emphasize how flexible we need be in adapting messages to overall learning preferences of audiences. The task seems to be in discovering learning styles through the use of planned feedback. Identifying a group preferred learning preference and incorporating that preference into the communication strategy pays off in higher retention and impact.

Credit: Multiple Intelligences and Learning Styles, Clipart 'Education' Word 2009 Office.

References and Further Reading

Banks JA (2007) An introduction to multicultural education, 4th edn. Allyn & Bacon, New York

Bennett CI (2006) Comprehensive multicultural education: theory and practice, 6th edn. Allyn & Bacon, Needham Heights, MA

Berens LV, Nardi D (1999) The 16 personality types, descriptions for self-discovery. Telos, Huntington Beach, CA

Blanchard KH, Carlos JP, Randolph A (1999) The 3 keys to empowerment: release the power within people for astonishing results. Berrett-Koehler, San Francisco

Brodie R (1996) Virus of the mind: the new science of the meme. Hay House, New York

Byrnes A, Conte A, Gonnot JP, Larsson L, Schindlmayr T, Shepherd N, Walker S, Zarraluqui A (2007) From exclusion to equality, realizing the rights of persons with disabilities, handbook for parliamentarians on the Convention on the Rights of Persons with Disabilities and its optional protocol. United Nations, New York

Castaneda A, Gray T (1974) Processes in Multicultural Education. Educational Leadership 32

Chisholm W, May M (2009) Universal design for web applications. O'Reilly Media, Sebastopol, CA

Dawkins R (1976) The selfish gene, 3rd edn. Oxford University Press, Oxford, England

Elgin SH (1980) The gentle art of verbal self defense. Prentice Hall, Englewood Cliffs, NJ

Felder RM (2010) Learning styles. http://www4.ncsu.edu/unity/lockers/users/f/felder/public/Learning_Styles.html. Cited March 9, 2010

Gardener H (1993) Frames of mind: the theory of multiple intelligences, 10th edn. Basic Books, New York

Gardener H (2006) Multiple intelligences: new horizons in theory and practice. Basic Books, New York

Goldberg MJ (1999) The 9 ways of working: how to use the enneagram to discover your natural strengths and work more effectively. Marlowe, New York

Goleman D (2006) Social intelligence: the new science of human relationships. Random house, New York

Hudson R, Riso DR (2003) Discovering your personality type: the essential introduction to the enneagram. Mariner Books, Houghton

Jung CG (1976) Introduction to type, 2nd edn. Center for Applications of Psychological Type, Gainesville, FL

Jensen EP, Nickelson L (2008) Deeper learning: 7 powerful strategies for in-depth and longer-lasting learning. Corwin, Thousand Oaks, CA

Louv R (2005) Last child in the woods: saving our children from nature-deficit disorder. Algonquin Books, Chapel Hill, New York

Pearce H, Brees KK (2007) The complete idiot's guide to the power of the enneagram. Alpha Books, New York

Quenk NL (2009) Essentials of Myers–Briggs type indicator assessment (essentials of psychological assessment). Wiley, New York

Rosenberg WR (2003) Multiple intelligence theory: applications to biological education, Master's thesis. University of Northern Colorado, Colorado

Satir V (1964) Conjoint family therapy. Science and Behavior Books, Palo Alto, CA

Satir V (1972) Peoplemaking. Science and Behavior Books, Palo Alto, CA

Terregossa R, Englander F, Englander V (2009) The impact of learning styles on achievement in principles of microeconomics: a natural experiment. Coll Stud J 43(2):400–410

Werneck C (2005) Manual on disabilities and inclusive development for the media and social communications professionals. People school – communication for inclusion and World Bank, Washington

Chapter 12
Communicating Across Cultures

12.1 Introduction

'Culture' is one of those expansive terms used most always without explicit definition. Like 'freedom,' 'love,' and 'common sense,' culture might seem to be intuitively understandable. But, grasping its meaning gets mentally slippery when we attempt to saddle it with a cut-and-dry definition. Ideas contained in the concept of 'culture' include:

- Integration of a society's knowledge, beliefs, and behaviors
- Education so as to transmit what is important to new members of a nation
- Customary beliefs, social norms, and material artifacts of a group
- The body of shared beliefs, values, attitudes, opinions, and practices of an organization

Imparting culture can be thought of as a form of education, though then we are using another concept that is larger than any definition. Undergirding the upkeep of culture is communication. The integration and transmission of the intangibles that make up society must involve messages. Lots of them.

So, how can we consider culture and use insights we gain to communicate better? Clearly, communicators deal in big ideas. Lots of them. Thinking back to the communication model in Chapter 2, consider the concepts of encoding and decoding of information. Even when an audience has been well-characterized, there still exists some 'noise' because the audience cannot perfectly decode a message. Multicultural aspects can cause receivers of a message to perceive things differently than was intended by the communicator.

Culture and communication are interdependent. One cannot exist without the other. Culture provides the blueprint that determines the way an individual thinks, feels and behaves in society. Communication gets those points across. We are not born with culture engrained in us, but learn it through acculturation and socialization. It is manifested through societal institutions, daily habits of living and fulfillment of psychological needs. Breach a cultural norm and those around censure you. Do well by their standards and they reward you with praise and admiration.

For environmental communicators, cultures can be conceptualized as analogous to ecosystems in the way they are nested inside one another. Drawing boundaries around either cultures or ecosystems may be necessary but is also ultimately imperfect. Environmental communicators, who carry with them intimate understandings of both the human world of culture and the wider world of ecosystems, are unusually suited for linking these two worlds and for passing that linkage onto others. This quality of seeing the human world within the larger natural one is environmental sensitivity. Environmental sensitivity has been shown to be one of the best indicators of environmentally responsible behavior (Sia et al. 1985/1986).

Awareness that there is no monolithic culture and that individuals come from varied and diverse cultural backgrounds is essential if a focused message, through whatever mechanism, is to be delivered successfully. A helpful concept to keep in mind when considering the broad spectrum from which members of your audience come is microculture – those cultures within cultures within broader macrocultures in which we all exist. These cultural structures are emphasized in Fig. 12.1.

Understanding why another person is reacting differently to a situation than you is a wonderful way to reduce conflict and to smooth differences, and also to garner understanding of a reality other than your own. We all have our own realities and worldviews. Until we are exposed to other realities we may not be able to understand why some people cannot see our points of view. Many people conflict, not on content, but on the process of communication itself. Understanding some of the different ways people process information and work through situations will help smooth communication.

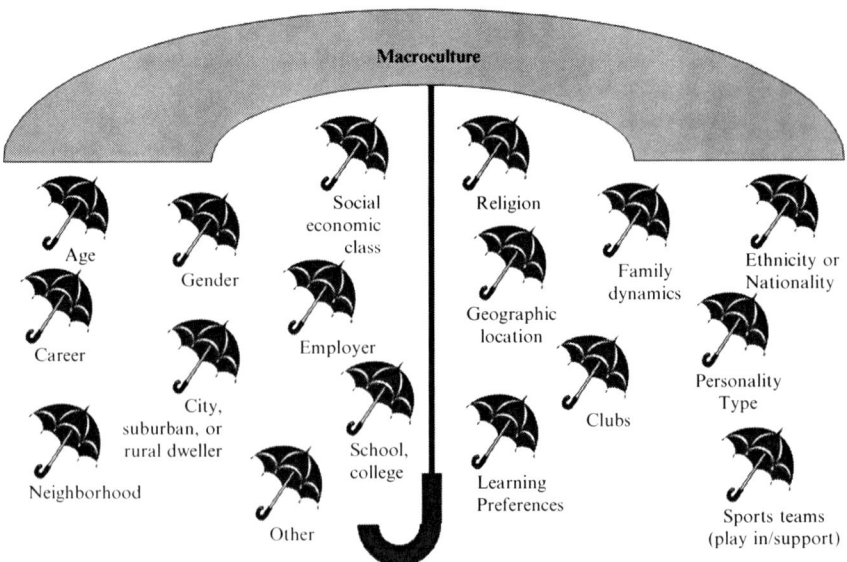

Fig. 12.1 Cultural identity umbrella model. Original graphic, Richard Jurin

Early childhood development defines who each of us are and what we accept as real, true and ideal. In order to understand others we need to see outside of our own personal boxes of reality. Exposing yourself to disconfirming information and other ways of being is a wonderful way to truly see your own particular set of cultures and to begin seeing other realities.

We all live and operate in many microcultures. But, only an individual who can successfully operate in two or more macrocultures can be termed bicultural or multicultural. Operating means to be able to work, understand and communicate easily within the culture where one is. The last two decades have seen rising value placed in multiculturalism. This is, perhaps, one corollary of the global village effect that became more and more evident in the last 50 years. Communicators are expected to be sensitive to the spectrum of cultural backgrounds of their message recipients.

12.2 Culture: Macro Versus Micro

Macroculture (also known as dominant culture) is the culture shared by most of a nation's residents. In addition to participating in the macroculture, each individual also belongs to a number of other cultures with patterns that may not be common to the macroculture. Using the United States as an example, within the dominant American culture one can easily recognize regional, ethnic and socio-economic variations. Think of Southern culture with its cuisine of sweetened iced tea, fried chicken, and grits. Or, Hispanic/Latin culture with its low-riders, burrito wagons, and rhythmic music. Compare those with Midwestern agrarian culture: the culture of hot dishes, working by the seasons, and family reunions where no one has to cross a county line to attend. And, upper crust blue bloods in eastern suburbs, with their inherited wealth, country club memberships and immaculate mansions. America has been called a melting pot, though many argue it is more like a tossed salad. Cultural differences within the macroculture make either metaphor accurate. An individual belongs to many different microcultures.

All those life ways in which you participate locally and regularly are your microcultures. Cultural identity occurs through the beliefs, traits and values learned through membership in microcultures based on national/ethnic origins, religious affiliation, gender, age, socioeconomic level, primary language/dialect, geographic region, place of residence (e.g., rural, suburban, urban), and other factors. The interaction of various microcultures within the larger domain of macroculture determines an individual's cultural identity. Membership in one microculture usually influences characteristics, beliefs and values of membership in other microcultures. Therefore, an individual is unique, but can still be defined by some characteristics associated with a specific microculture to which she may be a member. The importance of each microculture will vary between individuals even if they share similar memberships (Fig. 12.2).

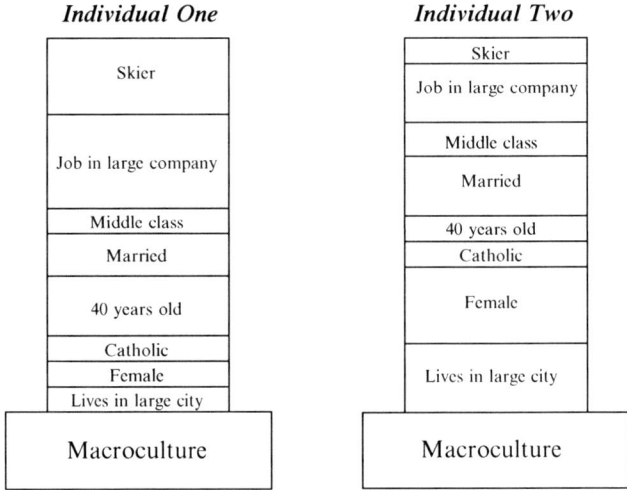

Fig. 12.2 Macroculture versus microculture model. Original graphic, Richard Jurin

12.3 Cultural Adaptation Theories

Individuals are exposed, literally from the moment of birth, to the details and telling aspects of their particular cultural milieu. This does not mean that there is no choice and role of free agency in this cultural education. But, to learn a culture one must have some exposure to it.

The process of learning cultures has been looked at and theorized on by hundreds of scholars. We'd like to look at several of the more accepted theories of cultural adaptation. It is not our purpose here to debate the strengths and weaknesses of any of these theories and especially how they might be, or might have been, applied to individual cases. Rather, we merely want to introduce them, so that you have some tools for looking at 'different' groups.

Assimilation – This is a process in which microcultures may be adopted into the macroculture. The process behind this concept is that there should be no competing dominant culture outside of the macroculture.

- **Conformity Theory** – This concept assumes the basic tenet of the dominant society is to encourage adoption (forcibly or voluntarily) of the macroculture to the exclusion of all microcultures.
- **Melting Pot Theory** – Emphasizes evolution of the macroculture through a mixing of culturally unique groups, to form a new culture expressing traits from all the cultures.

Suppression – In a suppressive situation, a microculture is deliberately isolated from the rest of the macroculture (by force or by choice). While the people in this isolated culture may develop a 'dualism' of identity, they are often viewed by the rest of the macroculture as inferior, and are often tightly controlled and/or sanctioned by all aspects of the society they live within.

Pluralism – This theory stipulates that cultural groups, particularly ethnic groups, maintain separate and distinctive identities from the dominant culture. This can be the idealized multicultural situation which operates through mutual respect and equity of all cultural groups.

Using these ideas to analyze and understand an audience informs message-creation for the communicator who takes time to inspect cultural nuances. Beginning to know the different microcultures from their particular perspective is helpful in forming an effective message that will be heard, understood, and acted on. One has only to think of the many minority groups in the United States, or any other county over the last two centuries, to begin realizing how these theories have all been applied at some time or other to some microculture.

12.4 Worldviews

A worldview is usually associated with an individual's concept of reality, but is also integral with a culture's collective view of itself. This is particularly true of how cultures relate to the natural world. Worldviews are derived from the cultural beliefs and values that a society forms over long periods of history. These factors are transferred from generation to generation and form the fabric of how people perceive their environment and determine what is supposed to be important. While worldviews may change slowly over time, understanding why a culture has developed specific traits of behavior will aid a communicator in formulating a message that will be decoded accurately as it was encoded.

It is often noted that large portions of world populations and many cultures share many, though not all, characteristics. Sometimes, being aware of a worldview can be the key between a successful audience analysis and communication, and a failed campaign. Many worldviews can be equated with Hall's concept of culture that sets cultures into High context and Low context (Hall 1976). Low context cultures tend to rely primarily on verbal messages with low emphasis on non-verbals. High context cultures have the opposite emphasis. A communicator should understand what type of cultural context is prevalent with any group that is being targeted with a message. Knowing the cultural context will help focus the quality of the message for a given targeted audience based on knowing what context is important for that audience.

Table 12.1, condensed and adapted from Bennett (1990), summarizes this concept:

Examples of how three different world cultures might differ through worldviews are (Hancock 1999):

Table 12.1 Worldviews from a high and low context perspective

	High context	Low context
Time	Polychronic (time not critical). Abstract view of time; loose schedules; lots of activity; quick changes	Monochronic (time of essence). Concrete need of time; tight schedules; linear use of time; stick to times
Space and tempo	High-sync. Valued harmony with others and nature	Low-sync. Isolation from others and nature acceptable
Reasoning	Comprehensive logic. Affective value; intuition, spiral logic, and contemplation of value	Linear logic. Cognitive value. Logical, analytical reasoning of value
Verbal messages	Restrictive codes. Reliance on nonverbal and contextual cues with 'shorthand speech.' In-depth, interpersonal, socially cohesive, and polite communication more important	Elaborate codes. Extended understanding of what is said or written with little reliance on cues. Little personalization with emphasis on argumentation and getting to the point
Social roles	Tight structure. Behavior predictable; role conformity expected	Loose structure. Behavior unpredictable; role conformity context-based
Interpersonal relations	Group is first. Well-defined and accepted status distinctions. More in-depth bonding and highly structured inter-relational dependency by all members of the group. Differentiation and/or mistrust of outsiders	Individual is first. Less defined status distinctions. Functional interrelationships within the group with little co-dependency. Less differentiation between members of group and outsiders
Social organization	Personalized law and authority. Customs and personal contacts are most important. Oral agreements are binding. When ordinary paths of action fail, contacts can 'bend' rules to affect action. Authority figures are completely responsible and liable for all actions by subordinates	Procedural law and authority. Procedures, laws and rules are most important. Written contracts are binding. Impersonal, often unyielding policy rules all paths of action. Authority figures negate responsibility when consequences are negative

Adapted from Bennett 1990.

12.4 Worldviews

African Worldview:

- Unity – everything is functionally connected.
- Oral tradition – oral history is important; language is participatory.
- Survival of the group/self-concept – individual finds identity in the 'We-ness' of the group; the group is primary.
- Extended kinship – each person is related to all other members of the tribe, ancestors, unborn, etc.
- Time – measured experientially; a focus on past and present; relaxed time.
- Perception of the environment – important to be in harmony with the environment.
- Activity – everything is continuous and connected; a sinusoidal view of activity.

Asian Worldview:

- Hierarchical authority – favorable view of authority; respect age, social status, teachers, parents, etc.
- Filial duty – respect and loyalty for authority figures is important.
- Collectivity – The group is more important than the individual.
- Self-concept – identity of the individual evolves through group membership, family, microculture, etc.
- *Gaman* – value is achieved through suffering and hard work.
- Social conformism – obedience to rules and regulations is highly valued.
- Activity – everything works together in harmony; a helical view of activity.

Euro-American Worldview:

- Competitiveness – be the best you can be.
- Time – future oriented.
- Belief in work ethic – value achievements and completion of tasks; visible and materialistic possessions define worth; delayed gratification.
- Individualism – individual effort valued over group effort.
- Dualism – body and mind held as separate; concrete concept of right and wrong.
- Perception of the environment – to be mastered and controlled.
- Self-perception – separate from the physical world; autonomy encouraged; solve one's own problems.
- Activity – isolate a problem, solve it, and get on with the next problem; a step-wise progression of activity.

Different microcultures derived from different ethnic origins will tend to incorporate the worldview of the original parent country, as can be seen in tight-knit immigrant enclaves in large cities. Others will have kept their original worldviews even under the overshadowing macroculture, as is the case with many indigenous peoples. The worldview not only determines how these cultures look at the world but how they think and even feel.

12.5 Where Are We Headed?

'We unconsciously and uncritically take our worldview for granted as the way things [always] are,' and always will be (Olsen et al. 1992). Most people rarely consider the historical influences that created the worldview they now have, and the fact that it is probably derived from earlier worldviews that have evolved to meet changing social and ecological situations. LaFreniere (1985) emphasizes, 'Changes in values and behavior related to nature would require the jettisoning of certain existing attitudes based upon beliefs about physical and metaphysical reality.' Leaving behind cultural fixtures born of hundreds of years of history will only be achieved when we consciously realize how they form a part of our current worldviews.

It has been proposed that industrialized societies are in transition between two worldviews, a historic *dominant social paradigm* and a modernistic *new environmental paradigm* (Dunlap et al. 2000). Jurin and Hutchinson (2005) support this idea showing that there does indeed seem to be a transitional worldview occurring. They name this worldview *logical idealism* and purport it to be as large as 40% of the population. Logical idealists show substantial pro-environmental leanings, but also internal conflicts about what they want in a lifestyle. This could be viewed more as a conflict between an espoused lifestyle of environmentally sustainable practices with respect for natural systems and a disdain for a highly wasteful consumerist way of life, which they see as a necessary evil. The need to maintain a first-class lifestyle is prevalent in this group. Communicating for ecological sustainability while emphasizing an increased quality of life with a high standard of living might yield more success toward achieving change towards a sustainable paradigm.

12.6 Empathy or Apathy?

Empathy, the capacity to recognize and understand another's state of mind and emotional state, is a competency within social intelligence. We do not tend to think about forces that act against the development of empathy. Modern consumerism prevents people who benefit from global capitalism from empathizing with those who are the burden-bearers of global environmental and social problems. They rarely think what it might be like to live in abject poverty while local resources are shipped elsewhere to propagate other people's monetary wealth. If more developed industrialized countries were genuine in their attempts to empathize the need for global-scale social change would be self-evident. Catastrophes like Hurricane Katrina in 2005 or the 2004 Indian Ocean Tsunami highlighted the plight of these most disadvantaged people, yet the speed at which these disasters were forgotten is an unfortunate indication of how little empathy truly exists.

Empathy is the pathway towards *compassion* and provides a compass towards deeper truths of human experience. Compassion is a profound human emotion that gives rise to an active desire to alleviate suffering of others. More than simply the recognition of another's suffering, empathy is actually sharing another's suffering, if only briefly. Longer lasting empathy develops into compassion. Empathy without compassion is merely *sympathy,* which is a social affinity in which one person stands with another person, closely understanding his or her feelings, usually for a short time before moving on.

The opposite of empathy are the dark twins, apathy and antipathy. *Apathy* is a state of indifference, an absence of interest or concern to certain aspects of emotional, social or physical life. It is a common reaction to stress where it manifests as 'learned helplessness.' *Antipathy* is a palpable dislike for something or somebody. While often induced by previous experience, it can also exist without any rational cause-and-effect explanation. It is often seen in unusual and stressful situations, and especially when people of different cultures interact. Strangeness or difference alone can create barriers to understanding. More often it is born out of a worldview gained within one's cultural upbringing. Consequences of apathy and especially antipathy are the raising of mental barriers to even the best-laid communication plans.

12.7 Stereotyping Versus Sociotyping

A term often used when talking about cultures is stereotyping. It is often used in a deprecating way. But, communicators must characterize their audiences to successfully target their messages. Here we define the terms of stereotyping and closely related subtyping and sociotyping.

Stereotyping – Psychologists have demonstrated that categorization by group, or stereotyping, is a natural phenomenon that humans use to develop mental categories that help make sense of the environment. We all stereotype because it is not feasible for our brains to process all the information continually bombarding us from our environment without some form of mental cataloguing. There is also a natural tendency for each of us to simplify our problems and to solve them with as much cognitive ease as possible. To analyze the behaviors, beliefs and values of every individual with whom we might interact would be an extensive time-consuming process (Bruner et al. 1986).

Stereotyping helps develop a sense of self through our connection with socialized groups. It helps us define who we are in relation to the rest of the world and in turn develops our specific worldview. Such categorization of information also helps us in relating to our environment. When we come across certain symbols, behaviors and patterns of speech, we expect a whole series of behaviors to occur, based on past experiences. These expectations then determine our reaction to a situation without having to think about it all that consciously. The broader the categories we learn to use, though, the less likely our overall perceptions are to be correct. These mental categories can often become too simplistic and cause us have a distorted perception of the world. Still, our stereotypes assist in creation of our worldview (Triandis 1971).

Stereotyping itself is not really either positive or negative. But, when stereotypes are overly broad or used without occasionally checking their veracity, then they can become dangerous. Such misuse may cause misconceptions, exaggerations and inaccurate generalizations to be developed about the environment and those groups in it. Stereotypes used to describe all members of a group without the possibility of exception is prejudice, and this is at the heart of history's never-ending supply of ethnic and international tensions.

Subtyping – Many times we will have a fixed stereotype of a group, and then we will meet an exception that doesn't fit 'the mold.' Exceptions to a stereotype which are encountered are usually subtyped. This means they are categorized apart from the main stereotype, as a deviant member of that category. This seems to be a cognitive mechanism to help us maintain the original (and perhaps inaccurate) category. If, however, we come across enough exceptions, we may begin to change that stereotype. Stereotypes are resistant to change. They usually only do so with much conscious thinking over a long time, a process not without cognitive dissonance. Recognize that we do stereotype and then to consciously suspend judgment about the categorization we form. In making no immediate judgments we can then see the group as a collection of unique individuals sharing some common theme.

Sociotyping – Mental categorizing is a natural phenomenon. In audience analysis, it is necessary to define a group so that a message may be correctly focused with the needs of the sender and the interests of the receiver. A sociotype is another form of categorization, that is more short-lived than a stereotype. It is usually a more accurate generalization about a social group or microculture. Rather than placing a value on individuals having membership within a group, sociotyping just describes the characteristics common to a group. In a sociotype, an assumption is made about the relationship of a specific group and a chosen attribute. But, not everyone in the group is assumed to exhibit this attribute. Sociotyping is therefore concerned with what characteristics individuals share within a defined group, and not how that group defines an individual. A sociotype is a conclusion reached by a communicator about an audience after analyzing it. Once the message for that audience is transmitted, the sociotype is allowed to dissipate. By not seeking permanence, sociotyping avoids the dangers of stereotyping.

12.8 Sensitivity for People with Disabilities

When interacting with people with disabilities, the sensitive communicator recognizes their needs, without being condescending or paternal. 'The challenge is to recognize that [disabled people] are more like [others] than they are different; to treat each as an individual; and to help them all achieve the greatest possible independence and personal, social and academic development' (Weisenstein and Pelz 1986).

To share messages with people with disabilities, you need to:

- Treat them equitably.
- Analyze the situation and their needs.
- Select the most appropriate media and instructional technique for the situation.
- Adapt the technique to make it most suitable for what you are trying to do.
- Be patience and tolerant.
- In group sessions, do not embarrass them by calling attention to their disabilities.
- Do not fill-in words for those who stutter.

Sensitivity should be shown to all people, but some special empathy may be given to people with disabilities. The key here is to employ empathy and not sympathy. People with disabilities usually have well-defined goals and understand their limitations (Weisenstein and Pelz 1986).

12.9 Cultural Awareness/Sensitivity/Competency

Globalization and international travel have increased the movement of people across the globe in general. It has become more crucial than ever that we all become more competent in understanding and learning about others cultures and improving our own cross-cultural capacity, regardless of where we come from. Working across cultures is a two-way street fraught with misconceptions and assumptions by everyone.

The terms cultural knowledge, awareness, sensitivity and competency have been used interchangeably, yet they are distinct. Developing these cultural understandings means recognizing the value of human diversity, regardless of cultural background.

Cultural Knowledge is to become familiar with selected cultural characteristics, such as knowing a different culture's history, values, belief systems and general behaviors, as distinct from one's own. Yet, knowledge of another culture is not enough to overcome barriers of misconceptions and misunderstandings.

Cultural Awareness means to truly understand another cultural group beyond mere recitation of facts, and involves internal changes in terms of attitudes and values. Awareness means to be more open and flexible in developing a relationship with cultural others.

Cultural Sensitivity is another step forward in knowing that cultural differences as well as similarities exist, without assigning values to those differences. Neither culture is seen as better or worse, as right or wrong.

Cultural Competency means to internalize knowledge, awareness and sensitivity about other cultures to such a degree that one is able to work effectively with individuals from different cultural and ethnic backgrounds, or in settings where several cultures coexist. Cultural competency includes the ability to understand language, social norms and behaviors of others, and to initiate appropriate interactions. Cultural competence exists on a continuum from incompetence to proficiency.

12.10 Becoming Culturally Competent

Just as empathy needs to be learned and nurtured, so too with cultural competence. While workshops and classes can increase knowledge, awareness and sensitivity, it takes a concerted desire and long-term commitment to become familiar enough with other cultures to deliver effective communications. This is especially true when dealing in environmental justice issues within the same culture as well environmental issues internationally. Communicators can work with different cultural groups by developing awareness, acquiring knowledge, and developing and maintaining cross-cultural skills. This includes understanding how other cultures view a communicator's culture. The difference in relevant cultural norms can be quite a culture shock for those not well-versed in other cultures. This is true for pluralistic countries where many people have been schooled in monocultural traditions resembling the dominant societal norms of those countries. If you have any doubts to your cultural awareness, consider how you treat cultural holidays, and then consider if other cultures would even know the deeper cultural significance of those holidays.

Steps to achieving cultural competency and maintaining cross-cultural skills can be summarized as:

- Admitting personal biases, stereotypes and prejudices, overcoming fears to these ideas, and engaging in continual evaluation of one's personal feelings and reactions to other cultural perspectives. This includes recognizing one's comfort, or discomfort, in cross-cultural situations.
- Valuing diversity by becoming more aware of other cultural norms, attitudes and beliefs. This will entail understanding how one's own culture is viewed by others. This can be done by personal immersion in another culture, or initially by attending classes, workshops and seminars about other cultures, by reading, watching culturally different movies and documentaries about other cultures, or by attending cultural events and festivals.
- Traveling to other cultural centers locally and then internationally. Be willing to make friends and establish working relationships with people of different cultures. Be willing to extend oneself psychologically and physically to another culture's norms.
- Showing respect for language by trying to learn another language, and become conversant with different cultural verbal and non-verbal cues.
- Being flexible and sensitive to others' feelings about their homeland. Developing nations are not as poor, backward or uneducated as Westernized countries seem to think. The key ingredient to developing and maintaining a long-term relationship with other cultural individuals is old-fashioned friendship built of mutual respect and a desire for understanding.

Communicating with a range of emotional and social intelligences is especially pertinent when learning to become culturally competent.

12.10.1 The Platinum Rule

A heuristic which can be fruitfully applied in cross-cultural situations is the *Platinum Rule:* 'Do unto others as they want done to them.' Based on the Biblical Golden Rule of 'Do unto others what you would like to be done to you,' the Platinum Rule recognizes differences between people that the Golden Rule ignores. Mindful use of the Platinum Rule will entail getting to know others before descending upon them with campaigns of social change. Bennett (1998) argues that practice of the Golden Rule in intercultural communications promotes sympathy, whereas the Platinum Rule encourages empathy.

12.10.2 Why Do We Have Cultural Conflicts?

Most cultural conflicts arise, not deliberately, but rather out of ignorance and a lack of sensitivity. A lack of empathy is also highly prevalent. Assumptions can be one of the most debilitating pieces of noise in any communications effort. A good communicator will make an effort to understand their audience as more than a target for a message, but as a group of people with similar hopes and expectations for the same kind of respect as they themselves would wish. This is as true for rich as poor communities, for less-developed countries as more-developed countries.

12.11 Conclusion

What is at the heart of this chapter is the need to be tolerant and understanding of differences we all have. These differences are based on cultural and social groups in which we are members. Depending on the culture into which we were born and raised, we will have certain ways of behaving, and also expectations about how things should be done. While categorizing is a normal part of thinking, we need to resist placing values on those categories to the exclusion of exceptions and evidence. For a communicator, it is paramount to understand the nuances of a culture into which you will place your messages. This will ensure correct encoding of a message and easy decoding, so as not to offend the target audience.

When using mass media, special emphasis should be given to cultural sensitivity. For instance, there are many gender-biased terms in use today. Terms like 'mankind' can be easily replaced by 'humans' or 'people.' Likewise, when referring to different ethnicities, potentially derogatory terms must be identified and substitute terms used. Insulting terms, phrases and deliveries cue your audience to turn away.

12.12 Case Study: Communicating Across Cultures. Cultural Context at Work

People in low-context cultures seem to accept work criticism more readily than those in high-context cultures (Earley et al. 2006). Low-context critiques work better when focused on outcome or process and not the person accomplishing the task. For example, a first draft of a pamphlet can be corrected for typos and graphic design without referring to the pamphlet creator's abilities. In high-context cultures, however, one must recognize the effort placed into creating a document before critiquing it for modifications and/or corrections. Saving face is a crucial component for the final success in the workplace. In many high-context Eastern cultures (and people from those cultures), the creator of a pamphlet might be insulted if a low-context culture supervisor directly critiqued the qualities of a draft. Feedback would need to be indirect to the pamphlet's 'errors,' but more direct about praise for effort already put into the piece. Rather than saying, 'The format needs to be changed to these specifications for this organization and clean up all the typos I've marked in red,' the culturally competent supervisor might say, 'This is a great draft pamphlet. Have you thought about doing it this way? There is a certain format this organization requires for this kind of medium.' In the latter case, the creator of the pamphlet is praised for their effort, the formatting emphasis is placed on the organization, not the individual, and the typos are not specifically pointed out to their face although they should be marked in the pamphlet.

Credit: Cultural Context at Work, Clipart 'Business.' Word 2009 Office.

References and Further Reading

Adler NJ (1996) International dimensions of organizational behavior, 3rd edn. South-Western, Cincinnati, OH
Banks JA, McGee Banks CA (2006) Multicultural education: Issues and perspewctives, 6th edn. Wiley, New York
Bennett CI (1990) Comprehensive multicultural education: theory and practice, 2nd edn. Allyn & Bacon, Boston, pp 55–56
Bennett CI (2006) Comprehensive multicultural education: Theory and practice, 5th edn. Allyn & Bacon, Boston
Bennett M (1998) Overcoming the golden rule: sympathy and empathy. In: Bennett M (ed) Basic concepts of intercultural communication, selected readings. Intercultural Press, Boston
Bruner JS, Goodnow J, Austin GA (1986) A study of thinking. Wiley, New York
Dunlap RE, Van Liere KD, Mertig AG, Jones RE (2000) Measuring endorsement of the new ecological paradigm: A revised NEP scale. J Soc Issues 56:425–442
Earley PC, Ang S, Tan J (2006) Developing cultural intelligence at work. Stanford Business Books, Stanford

References and Further Reading

Gollnick DM, Chinn PC (2008) Multicultural education in a pluralistic society, 8th edn. Allyn & Bacon, Needham Heights, MA
Gupta SR (2007) A quick guide to cultural competency. Gupta Consulting Group, California
Hall ET (1976) Beyond culture. Anchor Books, Garden City
Hancock C (1999) Personal notes on worldviews. The Ohio State University, Columbus, OH
Hark L, DeLisser H (2009) Achieving cultural competency: A case-based approach to training health professionals. Wiley-Blackwell, Malden, MA
Jurin RR, Hutchinson S (2005) Worldviews in transition: Using ecological autobiographies to explore students' worldviews. EER 11(5):485–501
Kenton SB, Valentine D (1996) Crosstalk: Communicating in a multicultural workplace. Prentice Hall, Upper Saddle River, NJ
LaFreniere GF (1985) Worldviews and environmental ethics. Environmental Review 9(4): 307–322
Lynch J (1986) Multicultural education: Principles and practice. Routledge & Kegan, London
Lynch J (1989) Multicultural education in a global society. Taylor & Francis, Bristol, PA
Olsen ME, Lodwick DG, Dunlap RE (1992) Viewing the world ecologically. Westview, Boulder, CO
Peterson B (2004) Cultural intelligence: A guide to working with people from other cultures. Intercultural Press, Yarmouth, ME
Schultz F (ed) (2000) Multicultural education 99/00, 8th edn. Dushkin/McGraw Hill, Guilford, CT
Sia AP, Hungerford HR, Tomera AN (1985/1986) Selected predictors of responsible environmental analysis: an analysis. JEE 17(2):31–40
Seelye-James A, Ned S, Knudsen A (eds) (1994) Culture clash: managing in a multicultural world. NTC Publishing Group, Lincolnwood, IL
Tayeb MH (1998) The management of a multicultural workforce. John Wiley & Son Ltd, New York, NY
Thomas DC, Inkson K (2004) Cultural intelligence: People skills for global business. Berrett-Koehler, San Francisco, CA
Triandis HC (1971) Attitude and attitude change. Wiley, Toronto
Urech E (2005) Speaking globally: effective presentations across international and cultural boundaries, 2nd edn. Smith-Kerr, Kittery Point, ME
Weisenstein GR, Pelz R (1986) Administrator's desk reference on special education. Aspen Publications, Rockville, MD

Chapter 13
Speaking to an Audience

13.1 Introduction

Most environmental and natural resource jobs require both the ability to write clearly and to speak well. While writing skills are practiced from our earliest grade school, public speaking usually gets only scant attention in our academic development. This chapter is designed to help you become a polished speaker. Specifically, three aspects of public speaking will be covered: structuring the presentation, delivering the presentation, and overcoming anxiety when speaking in front of an audience.

13.2 Structuring the Presentation

A superior presentation is well-structured, with an introduction, main body and conclusion. Main ideas are then more likely to be received and remembered by the audience. Just like a written document, a good presentation is created by developing an outline and working through several drafts. Preparation and practice help ensure the presentation flows well, and helps to reduce anxiety many of us feel about speaking in front of groups. Even if you know your subject intimately, the presentation must be developed for the specific audience in order to be effective. As always, know who the audience is and why this group is there to listen to you. How much time do you have? Do you need to leave time for questions from the audience? What do you want to achieve with this presentation? Where will you speak? Are the facilities suitable for what you wish to do? Being prepared and knowing the audience, room and setting helps reduce stress and helps you to become comfortable with speaking before people.

13.2.1 Interpretive Theming

When developing your presentation outline, it helps to think of a *theme* on which to hang your work, even if you are covering a seemingly mundane subject. A theme is not a topic, it is one or more key ideas in a presentation expressed in a meaningful sentence (Ham 1992). You are trying to make a cognitive and emotional connection of your message with the audience. A meaningful theme organizes, captures and sustains the attention of the audience. It provides a platform for the audience to consider, react to, build upon, appropriate and transform, thus allowing the audience to make their own connections to meanings within the message. An interpretive theme articulates a reason for caring about the message.

Interpretive themes are most powerful when they connect a concrete concept to a universal concept. A universal concept is an intangible meaning that has significance to almost everyone, but may not mean exactly the same thing to any two people. Universal concepts are ideas, values, challenges, relationships, needs and emotions that speak fundamentally to the human condition. For instance, standard of living is a tangible concept and sustainability is a universal concept. More simply, so are thirst and laughter.

Your theme should be stated on your outline as a short, simple, yet complete sentence, conveying one main idea. A theme reveals something insightful about the topic being covered and be compellingly worded when possible. If the theme does not draw an insightful and emotion response when first written, redraft it until you sense an emotional charge you would like your audience to have from the theme. It needs to say something important and powerful, so as to evoke and facilitate personal connections with the audience. A theme is more than just information. Remember, you can enhance the theme with your personal power during a presentation, but a good theme sets the stage for the rest of your presentation.

While you begin with a topic, the theme should give the topic some focus and punch. Examples of some topics and possible themes are (Table 13.1):

Table 13.1 Examples of topics and related thematic statements

Topic/subject	Thematic statement
Caves of Kentucky	Exploring caves is a sensual experience. Hanging out with bats preserves biodiversity.
Seasonal wildflowers	We manage our habitats to benefit both people and wildlife. Backyard wildlife needs your help.
Pharmaceutics plants	Our forest has many plants that heal. First Peoples had a natural treasure trove of medicines.
Bird migration	Flying 11,000 mi. a year in search of food is for the birds. Going south for winter keeps you warm and fed.
Cooking with native plants	You can't starve in the wild. Shopping in the wild for an indigenous person was dining out.
American railroads	Steam engines changed our lives in three ways. The night lights from space show us the path of the railroads in America.
Frontier homes	Living in the Jennings homestead was full of daily challenges. Frontier pioneers had to be Jacks and Jills of all trades.

13.2.2 A Presentation's Introduction

Your introduction should grab the audience's attention and orient them to your subject. Some ways to gain their attention are:

- Ask a question (to be answered in the presentation).
- Refer to specific people in the audience (assuming you have some familiarity with them).
- Make a reference to a recent event familiar to the audience.
- Tell a humorous or dramatic story to set the scene and illustrate your theme.
- Use audio-visual aids (cartoons work well).

In the introduction, the main theme of the presentation should be stated, along with the main propositions that support the thesis. Often, the introduction includes an explicit statement of goals and objectives the speaker is trying to achieve. Many speakers use feedforward to overview the talk and build interest at the same time.

Finally, the introduction should set the stage for you, as the center of attention. If a previous speaker introduced you, thank and acknowledge this, otherwise introduce yourself. Let the audience know how long you will be speaking and whether you want questions during the talk so that the audience knows what is expected of them. Many speakers, because of nervousness, apologize to the audience for lack of preparation or lack of knowledge of the subject. Remember that if you have prepared, this is not necessary and only makes everyone more nervous.

13.2.3 A Presentation's Main Body

The body of your presentation should contain three to five points that support the thesis of the talk. Effective speakers do not read the text of the presentation, nor do they memorize the body of the presentation. You should know the outline of the body well enough that you can 'hold a conversation' with the audience while still following the outline. This will create a relaxed, natural presentation rather than a speech which sounds mechanical or phony.

Develop an expanded outline of the presentation to organize your thoughts. Then create a summarized outline sheet as a guide to organizing yourself and the visual aids you might use. If you cannot use visual aids, then keep your presentation outline simple with just a word or two to trigger your thoughts. It is important to keep the outline simple, so you do not have to hunt for your next point. Above all, resist the temptation to read from your notes. If you can, develop visual aids to act as your primary guide. Each visual should prompt you on what to talk about next.

13.2.4 A Presentation's Conclusion

Your conclusion should summarize the points of the presentation, reiterate the theme and provide closure for the audience. The summary includes a restatement of the thesis and/or objectives of the talk, the supporting points from the body of the presentation, and indicates the importance of these points. Closure can be achieved by choosing an interesting quote that summarizes the thesis, by referring to events to follow the presentation, by providing a challenge to the audience, or by referring back to the introduction. In the conclusion, the speaker should not add new material, apologize or keep talking. Be concise and clean. A polished introduction and conclusion will set the audience at ease and create a professional image for the speaker.

13.3 Delivering the Presentation

Delivery includes the words you use, the way you speak them, the movements and gestures you make with your body in support of your words, and the grace with which you put these parts together. Outstanding delivery is almost as memorable as a great speech's content. Poor delivery can undermine the best prepared of presentations.

13.3.1 Verbal Delivery

Verbal delivery is the characteristic we notice most in excellent speakers. While delivery styles will vary according to the topic and the individual speaker, some general rules apply:

- Focus on the subject matter, not yourself. Do not wander off onto other subjects. The biggest problem we have when speaking is to constantly evaluate our own performances, which is nearly always much more critical than anyone who is viewing us in the audience. The audience is there to hear what you have to say and, believe it or not, they actually want you to succeed.
- Practice in front of a mirror. If you are not looking at yourself most of the time, either because you are looking at your notes or because you are not making eye contact, then your style needs some work.
- Do not depend on reading a script. Instead use key words from an outline. It will make you look more professional.
- Know the material and sound confident (even if you don't initially feel so). By projecting confidence, you also project credibility. You will begin to feel more comfortable in front of the audience.

- Do not bluff when answering questions. If you do not know the answer, say so.
- When the subject has been covered, stop. Do not ramble on too long. If you have a time limit, adhere to it. Once you have talked over time, the audience is probably no longer listening and instead will be anticipating your conclusion. Do not disappoint them.
- Be polite, smile and display respect for the audience. It helps the audience and you relax.
- Make eye contact with individual members of the audience. Let them know you are including them in your delivery. If they feel that you are talking with them, they are more apt to listen.
- Show passion for your topic. Enthusiasm will more than make up for discrepancies you may exhibit in a presentation.
- Be aware of your time limitations. Be aware of your time limitations. Be aware of your time limitations.
- Two golden rules of public speaking:
  - The talk should be well thought-out and prepared.
  - Be realistic: there is no such thing as the 'perfect speech.'

Ethos relates to the speaker's credibility and the concept of 'ethics.' It covers the whole essence of what character you are trying to project. As a speaker, you have to maintain credibility for your message to be conveyed to your audience. This includes exuding a sense of confidence, knowledge about your subject, and empathy for the topic. Having an enthusiastic and sincere delivery keeps your audience interested. Finally, you should look professional.

The speaker's appearance is an important factor in a successful presentation. Any presenter should want to appear knowledgeable, credible, trustworthy and acceptable to the audience. As a rule, dress in clean and neat attire. This affirms your status as a serious professional. It can be a judgment call, but to dress casually can make you feel more friendly. Yet, it could also depreciate your credibility as a professional. Most people have a stereotypical image of what a knowledgeable professional should look like. Correct dress codes may appear 'old-fashioned,' but that is what a majority of people are concerned with – old-fashioned values. Looking as though you just stepped out of a week living wild in the woods might be functional for a specific interpretational talk in a campground, but otherwise might lead your audience to question your credibility.

13.3.1.1 Vocal Qualities

There are some basic qualities that define a well-projected voice. Even if one does not have wonderful natural qualities like actor James Earl Jones, you can still be a dynamic speaker who captures the attention of the audience. Aspects that are crucial to being a well-rounded and credible speaker are inflection and projection, which are determined by pitch and loudness.

Inflection is determined by the pitch (tones of higher emphasis) and pace (delivery rate) at which we speak. While we all will agree that a person who drones on in just one predominant tone (monotone) at a constant pace can be boring to listen to, we must also be aware that the use of tones that become cyclic can also become boring to the listener after a few minutes.

In American and many British English dialects, there are usually only three to five tones (pitch levels) used to emphasize semantics within the language. A practiced speaker will try to vary the 'rhythm' of the tones, slowing and speeding the pace of delivery a little, to emphasize the more important or critical points within the content. It is also useful to clearly emphasize key words. Low tones indicate negative connotations, whereas higher tones emphasize positive aspects and extra emphasis.

Think of saying 'good dog' and 'bad dog' to a pet. A dog only hears the tones. If 'good dog' is spoken with a lower tone at start and finish, the dog might not be too happy to see you. Similarly, if both words in 'bad dog' were said in higher tones, the dog thinks it is being praised. It is usual to drop to a lower tone at the end of sentences to indicate the end of a spoken clause. But, it can be more effective to use a lower tone at the end to indicate something negative, or a higher tone to indicate something positive. Read, without emphasis, the following sentences, 'It was announced that 50 American Elm trees had died from the blight. However, 200 more had recovered because of the new treatment.' Put a lower tone on 'blight' and a higher emphasis on 'treatment' and notice how differently the sentences become.

Projection deals with the loudness at which we speak, as well as your voice's ability to carry. We all know the difference between a whisper and a shout, but we can have difficulty making our voices heard in a larger room without shouting. In a small group meeting, it is fine to speak with regular conversation volume, but in a larger room or an auditorium, especially one without a public address system, it can become almost impossible to hear someone who is not projecting their voice. Shouting is not a viable choice, because it will strain the vocal cords and may deafen people up-close. Projection, or carrying power, eliminates the need for harsh or shrill vocals. In order to avoid strain on your vocal cords, you should determine your 'optimum pitch' before any presentation. This can vary slightly at different times of the day and even on different days, dependent on mood and fatigue, but it should be within one or two tones of your habitual pitch. Habitual and optimal pitches are determined as follow:

Read aloud a piece of text without any inflection, until you can maintain a monotone. This is your habitual pitch which you use in everyday conversation.

Speak down the scale of tones and find the lowest tone you can produce. Next, count the tones as you speak up through the scales until you reach the top note you can squeak out. Your optimum pitch is about the fourth or fifth from the bottom, and will feel the easiest one with which to speak in a regular voice while projecting loudest. If your optimal pitch is not close to your habitual pitch, then practice speaking until the two are closer. This will help you to reduce voice strain.

In any setting as a speaker, you need to speak so people farthest away from you can hear easily and clearly. Breathing correctly is another important step to projecting well. It should come through the lower trunk of the body. While breathing, firm up your stomach muscles. This is called abdominal-diaphragmatic breathing. Try to push your stomach out while you breathe. You might want to try and push your stomach out against your hands until you get used to this method of breathing. Practicing this will allow you to control your exhaled breathing through the control of your diaphragm. Now couple this controlled breathing with well-formed vowels, and clear and precise articulation while you speak. Learning to relax the jaw will also yield fuller, firmer and resonating tones.

As a way to practice and actually see your breathing power during projection, see how you can affect a candle while breathing out using diaphragmatic control (Fetzer 1984). Pick a short phrase with well-rounded tones (e.g., 'How are you?'). Speak in front of a candle in your usual voice and notice how the candle flickers with your breath. Position the candle at the limit of when you make the candle flicker in a regular voice. Next try the breathing control and repeat the phrase. Do you blow the candle out? You should be able to keep moving the candle back a little more until you can affect it without shouting. Now try the breathing control again, but practice retaining your breath for longer periods, eventually of up to 40 s or more. When you can maintain a long breath, begin articulating a paragraph of text and imagine projecting to the back of an auditorium.

13.3.1.2 Mannerisms and Posture

Posture and mannerisms project a message about how we feel. Memorable public speakers succeed at projecting confidence and credibility. This can be practiced and learned. First, know how you look in front of an audience. Have someone video you while you present. This should show you aspects of your 'body language' that may be sending the wrong message. Standing straight without slouching are all part of a confident posture. If you suffer from not knowing what to do with your hands, hold a pointer or put one hand in your pocket (have your pocket empty so you don't play with coins or keys), or behind your back. Find what works for you. Try to avoid both hands in your pockets or having your arms folded in front of you. This tends to project nervousness or even hostility as a more tensed up posture.

If you can and it feels comfortable, do not be afraid to walk slowly about the front of the audience. Again, go with what works for you. Getting away from a single spot on the floor can be a useful stress-buster. If you are using a podium or lectern, try not to grip it tightly or hide behind it. Practicing before friends will give you feedback to develop your confidence. Remember, the more you speak publically, the easier it will become. Box 13.1 emphasizes some characteristics of various types of speakers. Aim to be credible and comfortable.

> **Box 13.1 Types of Speakers**
>
> What kind of speaker do you want to be? Some common types of speakers are identified below; you have probably encountered them all. In the future, when you are attending a presentation, observe the speaker and categorize him or her. What can you learn about your own speaking style?
>
> **Shrinking Violet**
> This kind of speaker does not address the audience; the only thing that can clearly hear the presentation is the speaker's shoes since that is where the speaker spends most of the time looking This type of speaker has lost the audience after a few seconds. If the audience does not feel involved in the speech or is bored, they will not listen. The same is true for a speaker that uses the words 'Um, erhhh, urhm, ya know' frequently. It gets too hard to listen and concentrate.
>
> **Zealous Animator**
> This kind of person addresses walls, ceiling and assorted inanimate objects in the room, and they do so with great theatrical flair by gesticulating wildly. This has the same effect as a shrinking violet in losing the audience, for now the audience is more interested in the arm-waving antics than the content of the talk.
>
> **Fire-and-Brimstone Preacher**
> A high-energy approach using high speed delivery and high volume is used to present the topic. This person may also pound the podium or nearby table, leaving audiences shell-shocked. A preachy speaker talks 'at' the audience in an overbearing and authoritative way. While this may have some effect with evangelical-type speakers in the correct setting, it tends to alienate most people.
>
> **Credible Speaker**
> Picture a relaxed, assured, confident presentation by a person who seems to enjoy public speaking. This type of speaker appears to enjoy presenting and seems to be talking 'with' an audience in a conversational style rather than just giving a lecture.

13.3.1.3 Influencing Audience Emotions

Emotions we project during a presentation can be contagious. We are able to trigger a range of emotion in others because our brains are hard-wired for sociability (Goleman 2006). Daily, we unconsciously take our emotional luggage with us wherever we go. Emotional contagion is where the brains 'low-road' circuitry operating below our awareness creates fast reactions to external cues such as aggression or happiness. This reaction is a kind of emotional 'shoot first and ask questions later' response. Therefore, the emotions we exhibit are drivers of the kind of day we will have. Conscious thought upfront can elicit a 'high-road' response, where we rationalize external cues before we decide on a reaction.

How do you react when a car driver that cuts you off in traffic? Welcome to the mood-driver in your brain. If you are aware of your mood-driver, then you can rationalize that the person cutting you off had an emergency and hence you need not feel hostility and remain calm. If not, you can react without thinking, become furious and engage in road rage leaving you seething for the rest of the day. Throughout the day our mood-drive operates continuously. In varying situations, we can feel great because of a kind act or annoyed because of a selfish act toward us. Emotions feed on each other. Our own states, as well as the ones of people around us, continually influence each other. Being conscious of our moods and the moods of our daily contacts can be moderated. Such moderation is not easy at first, but you can learn to do it without having to become His Holiness the Dalai Lama.

In a presentation, we can influence the audience's emotions. Think of this not as manipulation in a negative way, but rather as a way to enhance a message. Remember themes strike at emotional attachments to things we hold dear. This further plays out with the many non-verbal cues we may present.

13.4 Overcoming Anxiety About Public Speaking

Most people can identify with the fear of talking before an audience. Much of this fear is about something going wrong and embarrassing you. In a survey of the American public, presenting before a group was the highest ranked fear, so pervasive that 41% were scared of nothing more than public speaking (Wallechinsky et al. 1977).

When giving a presentation, we can become so fearful of things going wrong that we tense up and make it a self-fulfilling prophesy. What is the worst that could possibly happen? You forget everything you wanted to say, your notes are blown away by a sudden howling wind, all of your visual aids suddenly evaporate, and all your clothes suddenly fall off revealing your nakedness in public! So, perhaps you can't quite remember what you wanted to say…stop worrying about it and say what you can. After all, if you were well-prepared you can always refer to your outline or use your visual aids. The rest is just unfounded fear. And, if the worst should actually happen, the audience will be sympathetic and understanding.

Stress and fear are a normal part of our lives and should not be construed as negative aspects of who we are. Still, fear can be either rational or irrational. Rational fears prevent us from stepping off cliffs and getting into a cage with a lion. Irrational fears, however, tend to be misconceptions built into our individual realities because of past experiences. Identifying our fears is a beginning step toward resolving them. Box 13.2 contains a self-test to evaluate your fear with respect to public speaking.

Box 13.2 Speaker Apprehension Test

This self-test (McCroskey 1982) will help you understand your strong and weak points, in dealing with apprehension toward public speaking.

Directions This questionnaire has 24 statements on how you feel about communicating with other people. For each statement assign yourself the number of points that best corresponds with how you feel. There are no right or wrong answers, only your attitudes.

1 = strongly agree, 2 = agree, 3 = undecided or don't know, 4 = disagree, 5 = strongly disagree.

Many statements seem similar, but measure different aspects. Work quickly, recording your first impression.

1. I dislike participating in group discussions.
2. Generally, I am comfortable while participating in group discussions.
3. I am tense and nervous while participating in group discussions.
4. I like to get involved in group discussions.
5. Engaging in a group discussion with new people makes me tense and nervous.
6. I am calm and relaxed while participating in group discussions.
7. Generally, I am nervous when I have to participate in a meeting.
8. Usually, I am calm and relaxed while participating in meetings.
9. I am very calm and relaxed when I am called upon to express an opinion at a meeting.
10. I am afraid to express myself at meetings.
11. Communicating at meetings usually makes me uncomfortable.
12. I am very relaxed while answering questions at a meeting.
13. While participating in a conversation with a new acquaintance, I feel very nervous.
14. I have no fear of speaking up in conversations.
15. Ordinarily, I am very tense and nervous in conversations.
16. Ordinarily, I am very calm and relaxed in conversations.
17. While conversing with a new acquaintance, I feel very relaxed.
18. I am afraid to speak up in conversations.
19. I have no fear of giving a speech.
20. Certain parts of my body feel very tense and rigid while I am giving a speech.
21. I feel very relaxed while giving a speech.
22. My thoughts become confused and jumbled when I am giving a speech.
23. I face with confidence the prospect of giving a speech.
24. While giving a speech, I get so nervous that I forget facts I really know.

(continued)

> **Box 13.2** (continued)
>
> **Scoring** Follow the instructions. This questionnaire has one total score and four sub-scores.
>
> **Subscores:**
>
> | *Group Discussions:* | *Meetings*: |
> |---|---|
> | Begin with 18 points then add scores for items 2, 4 and 6 And subtract scores for items 1, 3 and 5 | Begin with 18 points then add scores for items 8, 9 and 12 And subtract scores for items 7, 10 and 11 |
> | *Interpersonal Conversations:* | *Public Speaking*: |
> | Begin with 18 points then add scores for items 14, 16 and 17 And subtract scores for items 13, 15 and 18 | Begin with 18 points then add scores for items 19, 21 and 23 And subtract scores for items 20, 22 and 24. |
>
> Each subscore ranges from 6 to 30. The higher the score, the greater the apprehension. Any subscore above 18 indicates a degree of apprehension to reduce. To obtain your total score, add the four subscores together. A total score above 70 may indicate shyness which can be overcome by practice and positive thinking. Consider joining a speakers group for feedback to help overcome many of the fears of public speaking.
>
> Used with permission: James C McCroskey.

Whatever the source of our fears, the body reacts with a basic flight-or-fight mechanism and floods our systems with adrenaline. It is this hormone that makes us feel jittery and anxious, and gears the body to action. Unfortunately, with irrational fears, we tend to have no outlets. It is this non-action that creates the poor performance typical of bad presentations. Understanding our fears allows us to refocus that energy and use it to good effect rather than suppressing it and impairing ourselves.

The following section will give you some ideas on how this refocusing can be best achieved for you. No single technique is right for everyone. Develop your own style that lets you become confident and relaxed.

Five different fears have been identified in conjunction with public speaking. Understanding these fears, and tricks for overcoming them, is important to overcoming speaking anxiety (Beatty 1988).

1. **Novelty** – Be prepared mentally for new situations. Not being prepared for novelty, such as doing an electronic presentation when one is unfamiliar with computers, will make you anxious.
2. **Subordinate Status** – Sometimes one is weighted down with a gnawing feeling that you are inferior to the audience. Think positively about yourself and understand the audience is there to listen to you because they want to hear what you have to say.
3. **Conspicuousness** – Though your presentation may be all set, you also need to be prepared to be the center of attention. In a conversation with friends, you would not mind being the center of attention. Visualize your audience as just a larger group of friends who want to listen to what you have to say.
4. **Dissimilarity** – In some settings, you might feel you have nothing in common with your audience. Understand and analyze your audience. They are there to listen to you. If you have targeted your audience correctly, you will be giving a talk that they honestly want to hear.
5. **History** – You are more likely to be anxious if you have been nervous before when giving a presentation. Think of all your previous experiences as learning sessions that prepared you to be better. After all, none of us were able to walk as a baby without lots of practice. The more you talk in public, the more your confidence will grow.

Finally, relaxation is important to giving polished presentations. Box 13.3 lists several powerful ways to relax the body and mind.

Box 13.3 Relaxation Techniques

Many times our stress can be actively reduced and refocused. Stress can remain positive when we take time to relax after or even during a challenge. The techniques can be effective in reducing stress in your life.

Deep Breathing
A. This exercise is central to all relaxation methods. Inhale slowly (taking at least 5 s) through the nose and expand the lungs completely. Exhale slowly (for at least 5 s) through the mouth. Repeat several times.
B. Repeat A, but this time put your hands on your stomach and, as you inhale, expand your stomach and your lungs. Your hands should rise if you are doing it correctly. A good addition is to say the word 'calm' as you slowly exhale. Both of these methods can be done anywhere at any time.

Mind-Clearing and Imagery
A. This practice is also basic to many relaxation methods. Find a quiet place without distractions. Get in a comfortable position. Close your eyes and do deep breathing. Form a mental picture of a peaceful place or event.

(continued)

Box 13.3 (continued)

    Concentrate on that image while you breathe in a relaxed fashion. Stretch upon completion of the exercise.
B. Repeat A, but continue the image and allow the mind to run free. Use your imagination to take a mental vacation of your choice whenever you need to relax.
C. Daydreaming is a healthy and useful technique if done so it does not interfere with the task at hand. It should be done at a time set aside to relax.

Stretching
Stretching various parts of the body can cause those muscles to relax. Concentrate on the muscle groups that bother you most. Relax, breath deeply, and slowly stretch the muscles that are knotted.

Moving
Many times we will stand behind a podium, thereby creating a barrier between speaker and audience. A good speaker can still make the audience feel involved in the presentation and also control any nervousness by channeling it into proper projection and inflection. A nervous speaker, however, will make the podium a haven to hide behind. Nervous speakers also will tend to exhibit other features such as involuntarily shaking the podium while gripping the sides with white knuckles, fidgeting and jerky mannerisms, and shuffling from foot to foot.

    In many cases, it is extremely useful if the nervous speaker can channel nervous energy in another direction that gives them some control. One such technique, if space is not limited, is to walk around slowly and deliberately, with varied stops. It can also help in eye contact since, as the speaker turns, they will automatically scan the audience. Although not recommended in all cases, pacing does afford a way to channel energy into the feet and away from the rest of the body.

Positive Mental Attitude (PMA)
A positive and healthy approach to life is the best way to insure a continuation of the same, as well as reducing the chances of developing a stress-related illness. Positive thinking is essential. Tell yourself, 'I CAN' rather than defeating yourself with negatives before you begin. Do not say, 'I can't.' Remember, BE POSITIVE. Communicate and share your experiences with friends, family or members of a support group. 'Getting it out' reduces stress. If a negative stress occurs in your life, there are three basic options to deal with it. First, try to remove the root cause of the problem, so it no longer exists. Second, develop new attitudes or behaviors to cope with the problem in a more positive way. The third option is to quit what you were doing in order to eliminate that stress in your life. Be an optimist rather than a pessimist. See that glass as half-full rather than half-empty.

13.5 Case Study: Public Speaking About and for the Environment. Speaking of Earth: Environmental Speeches that Moved the World

Great speeches inspire and provoke the audience to action. Alon Tal, founder of the Israel Union for Environmental Defense and professor of environmental policy at Ben-Gurion University, brings together 20 of the greatest environmental speeches ever. His volume presents full transcripts, alongside brief biographies, synopses of the environmental issue each speech confronts, and notes on speaking styles and delivery. Included are such stellar examples as:

- Thor Heyerdahl, Norwegian ocean explorer, 'If man is to survive, the ocean is not dispensable,' 1972
- David Lange, New Zealand's prime minister, 'Nuclear weapons are morally indefensible,' 1985
- Chico Mendes, Amazonian rubber-tapper, 'The destruction of our rain forest affects not only the Brazilian people, but in fact all the people of the planet,' 1988
- Ken Saro-Wiwa, Nigerian community organizer, 'A deadly ecological war in which no blood is spilled but people die all the time,' 1993

For the next generation of such a collection, it would be wonderful if videos could be collected. Then, environmental communicators could see and hear historical public speaking.

Credit: Speaking of Earth: Speeches that Moved the World. Copyright © 2006 by Alon Tal. Reproduced by permission of Rutgers University Press.

References and Further Reading

Anholt RRH (2005) Dazzle 'em with style: the art of oral scientific presentation, 2nd edn. Academic, New York, NY

Arredondo L (1990) How to present like a pro!: getting people to see things your way. McGraw-Hill, New York, NY

Beatty M (1988) Situational and predispositional correlates of public speaking anxiety. Commun Educ 37:28–39

Beebe SA, Beebe S (2005) Public speaking: an audience-centered approach, 6th edn. Allyn & Bacon, Boston

Bell AH (2008) Butterflies be gone: a hands-on approach to sweat-proof public speaking. McGraw-Hill, New Delhi

Bender PU (1995) Secrets of power presentations. Firefly, Willowdale, ON

Berkley S (1999) Speak to influence: how to unlock the hidden power of your voice. Campbell Hall, Englewood Cliffs, NJ

Booher D (1994) Communicate with confidence: how to say it right the first time and every time. McGraw-Hill, Palmdale, CA

Condrill J, Bough B (1999) 101 ways to improve your communication skills instantly. Goalminds, Palmdale, CA

Decker B, Crisp MJ (eds) (1997) The art of communicating: achieving interpersonal impact in business. Crisp, Los Altos, CA

Esposito JE (2005) In the spotlight, overcome your fear of public speaking and performing. LLC

Fetzer S (1984) Improving your speaking voice. Winning with words. World Book Encyclopedia, pp 79–107

Goleman D (2006) Social intelligence: the new science of human relationships. Random House, New York

Ham SH (1992) Environmental interpretation: a practical guide for people with big ideas and small budgets. Fulcrum/North American Press, Golden, CO

Jaffe C (2006) Public speaking: concepts and skills for a diverse society, 5th edn. Wadsworth, Belmont, CA

Krasne MT (1997) Say it with confidence: overcoming the mental blocks that keep you from making great presentations & speeches. Diane Pub Co, Darby, PA

Maxey C, O'Connor KE (2006) Present like a pro: the field guide to mastering the art of business, professional, and public speaking. St. Martin's Griffin, New York

McCroskey JC (1982) Introduction to rhetorical communication, 4th edn. Prentice Hall, Englewood Cliffs, NJ

Mckerrow RE, Gronbeck BE, Ehninger D, Monroe AH (2000) Principles and types of speech, 14th edn. Longman, New York

Monarth H (2007) The confident speaker: best your nerves and communicate at your best in any situation. McGraw-Hill, New York

Reynold G (2008) Presentation zen: simple ideas on presentation design and delivery (voices that matter). New Riders Press, Gresham, OR

Rozakis LE (1999) The complete idiot's guide to public speaking. McMillan, London

Swets PW (1992) The art of talking so that people will listen: getting through to family, friends, and business associates. Prentice Hall, Englewood Cliffs, NJ

Tal A (2006) Speaking of earth: environmental speeches that moved the world. Rutgers University Press, Piscataway, NJ

Timm PR (1997) How to make winning presentations: 30 action tips for getting your ideas across with clarity and impact. Career Press, Franklin Lakes, NJ

Urech E (2005) Speaking globally: effective presentations across international and cultural boundaries, 2nd edn. Smith-Kerr, New York

Wallechinsky D, Wallace I, Wallace A (1977) The people's almanac presents the book of lists. William Morrow, New York

Wilder L (1999) 7 steps to fearless speaking. Wiley, New York

Woodall MK (1996) Thinking on your feet: how to communicate under pressure, 2nd edn. Professional Business Communications

Zarefsky D (1999) Public speaking: strategies for success, 2nd edn. Allyn & Bacon, Boston, MA

Chapter 14
Communicating Without Words

14.1 Introduction

When discussing communication, we intuitively think of the language of words. Words are powerful conveyers of meaning and they do matter in getting your message across. But, there is a whole other realm of communication, one that transfers meaning between sender and receiver without using words. This is the realm of nonverbal communication (also referred to as 'subtext').

It has been estimated that between 60% and 95% of the meaning transferred in a communication system is accomplished through nonverbals. This is more than 'body language.' Nonverbal communication includes physical and psychological signals, subtle inferred meanings in the motions of sender and receiver, implications of unspoken words, and buried cultural expectations present when communicating. Indeed, we often rely more on nonverbal cues than the actual words used. For example, a friend says, 'I studied all night' but stresses the word 'all,' laughs and rolls her eyes. You immediately know she is being sarcastic, and did not spend much time studying in reality.

It is not just what you say with words, but how you say them. Nonverbal communication is all the ways we add or change meanings of the words actually spoken.

In this chapter, we discuss the ways we are socialized in the use of nonverbal communication. We hope to make you aware of many nonverbal forms of communicating to help you understand your own patterns and those of people with whom you interact. Knowing nonverbal language, however, is not a reason to practice deception. It should be used to enhance your spoken message to improve clarity and increase credibility. It must be emphasized that nonverbal language is dependent on culture, since many symbols and movements can vary dependent on the user or receiver's experience and semi-conscious intent. All cultures exhibit nonverbal communication and often in a similar fashion.

14.2 Kinesics: Physical Movement

Whenever a person feels confused or senses a contradiction in a conversational setting, it is often because the nonverbals, from certain physical movements and body actions, are not matching the words being spoken. Kinesics (the study of gestures, body language and facial expressions, especially in combination with speech) can be separated into three aspects, though all three usually occur at the same time. You are often aware of this at a subconscious level, and so are often reacting at an emotional level to the kinesic messages you are receiving.

- **Body language** – It is estimated we use body language two-thirds of the time we are in personal communication, usually subconsciously. The dynamics of body language, a sequence of motions and gestures, lend meaning to the subtext. Examples of body language are leaning toward someone you find interesting, turning your shoulder and back toward a third person you are 'excluding' from a three-party conversation, and crossing your arms in front of you to create a barrier. Becoming aware of the movements can help you to communicate clearer. By matching your body language to your spoken language you will increase your credibility.
- **Gestures** – Gestures are dependent on the context and situation in which they occur. The same gesture may have several meanings at different times and places. Handshakes are the basic measure of respectful greeting in the United States. How you shake can affect how the other person feels about you. In Japan, the angle of the bow determines whether you will insult the other person or demean yourself, but this is usually waived for non-Japanese, who need only bow to show respect. A single wave of the back of a hand may be seen as an insulting dismissal or just a 'no problem' comment. Raised eyebrows can be a sign of surprise especially when accompanied with a smile, but one raised eyebrow with a serious expression can be a sign of disapproval or disagreement. In a boss-to-subordinate situation, the same gesture can become a power play, or at least diminish the confidence of the person speaking. Naturally, most cultures have some gestures that have a derogatory nature about them and should not be used in any professional circumstances, even when angry.
- **Facial Expressions** – including eye movements: Raising eyebrows or smiles can usually be friendly but can mean disrespect or aggression if held too long. In American culture, whenever someone holds eye contact for more than about one second something additional must happen – either a smile, a nod, a blink or a salutation of some form is customary.

14.3 Proxemics: Personal Space

We all have an individual zone of personal space which surrounds us. These zones are where we position ourselves in relationship to others. Therefore, how we situate ourselves is critical to our personal comfort, yet we need to be aware of how we impact other peoples' zones. This bubble seems to be larger in areas that we can see, especially extended to the front and somewhat at the sides. Examples where proxemics can be readily seen in action are in the elevator, how people arrange themselves on chairs in a waiting room, or where people sit at a bar. Take some time and watch how people arrange themselves when in groups in various locations and at different functions. The distances given below are based on pioneering research by Edward T. Hall (1963).

- **Intimate Zone**: 0–18 in.
  This zone is reserved for highly personal relationships. Being uninvited within the intimate zone can create a feeling of anxiety in the other person and lead to aggressive reactions. Think about a crowded elevator. Did you feel anxiety about people standing close to you? It will become easy to spot those people who are comfortable with each other, those that need their distance, or are strangers.
- **Personal Zone**: 18 in.–4 ft
  Most people in a professional situation will adopt the personal zone and maintain the distance. Try talking to someone and then slowly step back and to the side. The other person should follow your pattern in order to maintain the same distance. To talk from too far away is to appear cool and distant, and too close is to be invasive. We naturally feel our correct distance even though it is culturally learned. If your comfort distance is closer than someone else's, you will notice signs of discomfort in them. That is your cue to slightly increase the distance between yourselves.
- **Social Zone**: 4–12 ft
  In a group, people will tend to space themselves equally several feet from the speaker, yet close-up within the group. Since individuals in groups are usually standing side-to-side they do not require as much space as they would face-to-face. If the group is larger and feeling 'crushed' there will be much unconscious negotiation of space between each of the members of the group. The less comfortable members of the group may stand far to the back to increase their space or even leave the group. If this happens then the speaker might want to actively get the group into the public zone where each member can claim whatever space they need.
- **Public Zone**: 12 ft and more
  In this zone, the distance of the speaker is rarely a problem except that they are now farther from the audience and thus are in danger of losing that special connection that allows personability. The speaker needs to become more dynamic to retain a connection with the audience. As the group gets bigger, there is increased negotiation on space. Seating arrangements in a theater or community hall need to be carefully thought out to ensure everyone can have adequate space.

Be aware of the distance people need to be comfortable. If they are feeling anxious because of proxemics, they are probably not listening well.

14.4 Semiotics: The Science of Symbols

Semiotics can be thought of as application of nonverbal communication to influence an audience. Here, kinesics and proxemics are brought together, so nonverbals create a favorable impression on a listener. Semiotics could also be used to create a pleasant environment for an important meeting. Below are several ways of using semotics.

- **Physical Appearance** – You have probably heard the sayings 'neatness counts' and 'dress for success.' Appearance varies for different ethnic groups/cultures. While it may not be correct from a multicultural perspective to judge a person by their appearance, it would be naive to not believe that credibility, professionalism and even respect are gauged by the type of clothes that one may wear in any specific setting.
- **Smell/Taste** – Overlooked as nonverbal factors, they can have significant affects on moods and psychological associations. Perfumes and fragrances, for example, while pleasing in some situations like a romantic evening, can become quite distracting in a professional setting. Certain tastes and smells can evoke intense memories. In general, be careful when setting up a situation involving odors or tastes.
- **Eye Contact** – The way we look at people and how we use our eyes during personal interactions indicates a lot about what we are thinking. In Euro-American culture, it is usually expected to hold eye contact while talking to someone because this indicates that you are listening. But, eye contact between strangers becomes uncomfortable if held for more than a second or two. Some cultures use continued eye contact as a form of aggression (an invitation to fight) while others use it as a sign of warmth and respect. If you are involved with different cultures, it behooves you to learn the cultural nuances for your own benefit.
- **Smiles** – Smiling is believed to be universal in cultures all over the world (Konner 1987). This is one facial expression that has the capacity to warm up most any situation and enhance any presentation. It usually conveys warmth and positiveness. Of course, like any action, it can have a negative connotation as with 'the crocodile smile' where the person smiling is insincere. Usually a whole nonverbal review of the smiling person should reveal whether the smile is genuine or not.
- **Haptics** (physical touching) – If in doubt, keep your hands to yourself. If you use haptics in your everyday body language, become aware of how you use them and in what situations so as to use them for positive results. If you do not use haptics be careful of beginning them until you understand the full implications of what a touch means or how it can be misconstrued. In today's society, many haptics may be misconstrued by people as harassing or as plays for dominance. By all means do use, or learn to use, haptics for they are one of the most influential nonverbal factors to show caring and empathy. Frequency of touching is also a positive and strong aspect of haptics, if done suitably for a given situation.
- **Gestures of Respect** – These may be greeting gestures such as handshakes and bows. They also may be simple things like type of clothes. Wearing the right clothing for a specific situation (e.g. job interview) signals respect for the other person.

- **Note-Taking** – It complements the speaker because you appear to be interested in what they say. Be sure it is acceptable in the culture and under the situation.
- **Dominance Factors** – Be aware of aggressive power play. How you use nonverbals in a semiotic way can be either negative or positive. If you feel negativeness is occurring, try to assess your nonverbals or those of the people with whom you are communicating to decide if one of you is trying to be dominant over the other. If you are having a positive interaction, assess the nonverbals again to understand what you are doing that makes it so.
- **Setting the Scene** – Create an environment for success with the correct setting. Imagine trying to have a romantic dinner on a busy railway station platform! Obviously, a more remote and subdued lighting environment would be more aesthetic and appropriate. Similarly, think of having a business luncheon in a pleasant restaurant. It usually confers a less formal atmosphere and may germinate more open communication.
- **Seating at a Conference Table** – Who sits at the head of the table and who sits next to them? How is power identified just by seating arrangements? Are you trying to enforce your dominance or do you wish to create more equitability? If you create a seating arrangement of 'us' opposite 'them' then the table becomes a barrier. Preferential seating can also create enmity within the group. Or alternatively, establish the 'pecking order' and reduce potential conflict. As usual, knowing your audience is essential.

14.5 Paralanguage

There are several ways we use spoken language to give additional, and occasionally alternative, messages, other than the one we are actually saying. Such uses of paralanguage are:

Articulation – Speaking clearly and confidently increases credibility

Pronunciation – Helps an audience clearly understand what has been said, thus improving transference of meaning

Emphasis – Putting emphasis on certain words gives those words a different meaning and brings an audience's attention to those words. Ways that this often happens include:

- Pitch – Using a higher pitch for positiveness or a low pitch to express negativeness
- Rate – Speaking slowly to bring more emphasis to every word, or fast to gloss over certain words
- Timbre – Imposing a harshness or softness in the words, or whispering instead of shouting
- Pauses – This can give … dramatic effect … to the words just spoken

14.6 Psycholinguistics

When kinesics, proxemics, semiotics and paralanguage are composed together in a communicational situation, the communicator is using psycholinguistics. With psycholinguistics, ulterior meanings and dominance may be in play. Think of the following sentence: 'Jim has done a great job on this assignment.' Say this sentence three different ways, but use nonverbal techniques and voice adjustments to give a different meaning each time to the sentence. Think how voice inflection and nonverbal communication can alter the meaning of a message. If there is a conflict between the implied message and the verbal message, the implied (nonverbal) message is more likely to be believed.

Presupposition Statements: Elgin (1980) emphasized how the spoken word can be changed in many ways to uncover hidden meaning. Read through the list below and then think of the hidden presupposition that is represented. How often have you been a receiver of a negative presupposition that has left you angry and unsure of why you are so upset? Alternatively, do you use presuppositions as a form of your own and wonder why people are so easily upset by you? Some examples, derived from Elgin (1980), are given below. The word(s) to accent is bolded in each line.

- If you **really** cared, you wouldn't go ….

Presupposition: You don't care!

- If you **really** cared you wouldn't **want** to go ….

Presuppositions:

You don't care.
You have the power to control your decision.

- Don't you even **care** about your weight?

Presuppositions:

You don't care.
You should care, it's wrong not to.
You should feel guilty and rotten.

- Even a **man** should be able to understand **biology**.

Presuppositions:

There's something wrong with being a man.
It doesn't take much to understand biology.
You should feel stupid and guilty.

- A person who **really** wanted to save money would spend it **carefully**.

Presuppositions:

You don't want to save money.
You spend money carelessly.

Elgin recommends that in order to defend yourself successfully against negative presuppositions, you need to identify the presuppositions and then be sure to address them. In the last statement, you would need to address **both** your desire to want to save **and** your wish to be a careful spender. To ignore either means you have already lost the argument!

14.7 Metaphors

Metaphors substitute the name of one thing for another. Cultural analogies may imply the nature of something that is only known because of being a member of a specific culture. An example of this is the Navajo 'code talkers' in the Pacific theater of World War II. These coders spoke to each other in the Navajo language over open radio channels, yet the only people who could understand the message were other coders who knew the tribal myths and stories and could identify what was implied in the messages. Enemies were never able to crack this code, even those they were hearing the messages freely.

In any culture, there is a great degree of language that uses metaphors to set up concepts for receivers (Lakoff and Johnson 1980). If using metaphors with a specific audience, ensure they understand each metaphor's real meanings and do not read too little or too much into a metaphor being used. By virtue of metaphors being analogies, they are not exact and it is the extra parts that may create problems. As an example, a natural resources professional may make reference to a situation being a 'tragedy of the commons.' Audience members who are unfamiliar with the work of Garrett Hardin will not understand the point, however. So, while metaphors are useful to create analogies, they need to be based on shared understanding.

14.8 Cultural Implications

Nonverbal communication depends on the cultural context in which it is delivered. Differences may be ethnic, microcultural or macrocultural. Though picking up on subtleties can take years, here are some simple rules for using nonverbals within cultures other than your own:

- Keep body language simple to be less confusing.
- Know your audience when giving an oral presentation. Be sure your nonverbals and language mean the same thing to the audience that they do to you.
- Understand that nonverbals are dynamic (made up of a sequence of movements) and you should be wary of trying to interpret singular actions. Individuals may have developed unusual personal meanings for some actions, yet will still comply to cultural norms for overall body language.

- Develop a conscious sense of unspoken meanings around the words in a conversation. This is an ideal way to begin to reduce conflict and dominance factors, and thus avoid inadvertently offending or hurting people.
- Though there will be a sequence of nonverbals, we still subconsciously fixate on micro-expressions, tiny shifts in small facial muscles. Social psychological Paul Ekman (2009) has studied human abilities to tell lies and to deceive. The advantage of understanding micro-expressions, besides for making unusual TV cop shows (which Ekman has contributed to), is to help a communicator establish truthful and constructive personal communications. It is possible to train yourself to better recognize micro-expressions in an hour, claims Ekman (2009).

14.9 Consistency in Using Nonverbals

Our brains are hard-wired to consider and categorize external cues, though much of this mental processing takes place beneath the threshold of consciousness. So, you could contend most of our communications is done nonverbally and unconsciously. When we deliberately use nonverbals to influence, we need to be fully cognizant our verbal cues match our nonverbal cues. True enthusiasm and passion about our topics can naturally create such synergy.

The most straightforward and least deceitful way to use nonverbals is as honestly as you can – match your nonverbals to your spoken message. When using nonverbals during community capacity-building or dispute resolution, concentrate on the cues of others as well. Careful perception can help reveal hidden agendas that create barriers. Recognizing these can facilitate constructive processes and communication.

Yet, you may in certain situations feel someone is being insincere despite outward signals that indicate warmth and sincerity. Do not discount these perceptions, as you may be picking up on minimally perceptible nonverbal cues. Ekman (2009) calls these micro-expressions. Fortunately, it is hard to lie. Lying takes a highly concerted, conscious and continual effort. Attune yourself to detect liars.

If we adopt a positive humanist approach, 'forthrightness' is assumed to be our brains' natural response to a group challenge. Forthrightness will be mirrored by the body's nonverbal cues too. Through these cues, members of a group find their rapport with each other, leading to a feeling being 'in-tune.' When this happens for a group, interpersonal communications flow more freely and find successful resolutions quicker.

To enhance our environmental messages, how do we ensure a positive emotional context? To create rapport, three elements are needed:

- Mutual attention
- Shared positive feeling
- Well-coordinated nonverbal portions of one's messaging

14.10 Conclusion

Nonverbal communication is a critical yet often overlooked aspect of delivering a message. Consider the difficulty of holding a meeting between parties with little trust – perhaps between a community action group and a large corporation. What actions could be taken based on what you have learned in this chapter to create an atmosphere of trust?

14.11 Case Study: Nonverbal Communication. Nonverbals between Superior and Subordinate Workers

Two actors were used in a study of superior-subordinate interaction (Remland 1984). They both did a high-status (showing superiority through relaxed posture, indirect body orientation, loud voiced, inattentive behavior and spatial invasion) and a low-status (being subordinate with tense posture, direct body orientation, soft spoken, hesitating speech and attentive gaze) identical role-play interaction in which a superior was reprimanding a subordinate. The script was the same for each interaction with just the nonverbal cues changed to reflect either high-status or low-status interaction for each of the superior and subordinate interaction options. An audience was asked to rate the considerateness of the superior for each role play. The audience rated the superior more considerate when both superior and subordinate concurrently exhibited either low-status or high-status nonverbal. When status differences are decreased the communication is felt to be more equal and fair. Therefore, superiors will benefit by assessing a subordinate's nonverbal cues and matching the status being exhibited by the subordinate. This is especially true in cultural interactions and community interactions where 'power distance' is often prominent. It has been proposed that dysfunctional leadership may result from asymmetric superior-subordinate interactions that alienate subordinates, or in another perspective, experts projecting superior attitudes when working with communities asking for help.

Credit: Nonverbals between Superior and Subordinate Workers. Clipart 'Acting' Word 2009 Office.

14.12 Case Study: Proxemics. 'Personal Space' functional artwork by Vivian Puxian

Brazilian artist Puxian demonstrates her wearable art piece 'Personal Space' in Madison, Wisconsin, U.S., in July 2009. Her invention makes tangible the zone carried in public by each person. She toured the United States wearing her art in summer 2009, getting second looks, laughs and lots of discussion everywhere she went. In addition to the nonverbal communicative norms pointed out by her work, Puxian made this contention: 'It also protects you against infectious and contagious diseases, such as swine flu. Keeping a safe distance of 3 to 6 feet from infected individuals decreases the likelihood of contracting harmful germs.'

Credit: Personal Space "functional artwork by Vivian Puxxian. http://personalspaceprotection.blogspot.com

References and Further Reading

Breasure J (1982) Non verbal communication skills. Advanced Development Systems. Arlington, VA
Brownell A, Bache-Wiig T (2007) Non-adversarial communication: Speaking and listening from the heart. Velvet Spring, Boulder, CO
Dimitrius J, Mazzarella M (2008) Reading people: How to understand people and predict their behavior, anytime, anyplace. Ballantine, New York, NY
Ekman P (2007) Emotions revealed: Recognizing faces and feelings to improve communication and emotional life, 2nd edn. Holt Paperbacks, New York, NY
Ekman P (2009) Lie catching and micro expressions. In: Martin C (ed) The philosophy of deception. Oxford University Press, New York, NY
Elgin SH (1980) The gentle art of verbal self-defense. Dorset press
Elgin SH (2000) The gentle art of verbal self defense at work. Prentice Hall, New York, NY
Elgin SH (2009) The gentle art of verbal self defense. Fall River, New York, NY
Fast J (1994) Body language in the workplace. Penguin, New York, NY
Fast J (2002) Body language. M. Evans and Co, New York, NY
Goleman D (2006) Social intelligence: The new science of human relationships. Random House, New York, NY
Gorman CK (2008) The nonverbal advantage: Secrets and science of body language at work. Berrett-Koehler, San Francisco, CA
Hall ET (1963) Proxemics: A study of man's spatial relationship. In: Galdston, Man's image in medicine and anthropology, monograph IV. International Universities Press, New York, NY
Knapp ML, Hall JA (2009) Nonverbal communication in human interaction. Wadsworth, Belmont, CA
Konner M (1987) The enigmatic smile. Psych Today 21:42–46
Kovecses Z (2002) Metaphor: A practical introduction. Oxford University Press, Oxford and New York, NY
Lakoff G, Johnson M (1980) Metaphors we live by. University of Chicago Press, Chicago, IL
Lawley J, Tompkins P (2000) Metaphors in mind: Transformation through symbolic modelling. Developing Company, London
Navarro J, Karlins M (2008) What every BODY is saying: An ex-FBI agent's guide to speed-reading people. Collins Living, New York, NY
Pease B, Pease A (2006) The definitive book of body language. Bantam, New York, NY
Remland MS (1984) Leadership impressions and nonverbal communication in superior-subordinate interaction. Commun Quart 32:41–48

Chapter 15
Using Visual Aids

15.1 Introduction

Going far beyond mere words, communication through visuals enhances presentations. Graphic design is creative expression of messages using symbols, colors, shapes, text and the arrangement of all these elements. Environmental communications can enhance their own products with a few graphic design skills.

This chapter is intended to guide planning and presentation of visual aids. Visual aids, as explained here, are meant to boost a personal presentation, while a visual message is meant to present a stand-alone message in the absence of a presenter (e.g. a park interpretation sign or an advertisement). Actual construction of visual aids is covered in other texts, some of which are recommended at the end of the chapter. What sets off a visual aid from a visual message is how it is applied and used.

15.2 Visual Aid Basics

The old adage is that a picture is worth a thousand words. To be a visual aid, a valued picture need be the right picture, presented well, at the right time, or else it will detract from a presentation. A visual aid which is well-designed and suitably placed can give much more clarity to a message in a shorter time than a lengthy set of words. Combining both words and visual aids greatly increases impact. Visual aids are also important because of differences in ways people perceive messages. Some audience members will react best to verbal information, some to visual, and some to written material. Thus, presenting messages in multiple formats shows sensitivity to learning styles and multiple intelligences. In general, though, people tend to remember more of what they both see and hear. In addition, visual aids increase an audience's attention span, and can, when used properly, add structure to information to facilitate understanding.

There are many types of visual aids. Using the right one in a particular situation relies on an understanding of the strengths and limitations of each. In essence, anything that visually depicts a message either directly, subtly or semiotically is a visual aid. As a departure point, the presenter should draft the presentation first and then find visual aids that fit and enhance the presentation and not vice-versa. Remember, the message is what is important. The visual aids are just enhancements to help the audience get the message.

A key point when using visual aids can be summed up in the term 'Say dog, see dog.' If you are talking about a dog, then the visual aid being used should be a dog. Likewise, if the visual aid you are displaying is a dog, then you should be talking about a dog. This rule is often neglected in presentations. For example, if you get a puzzled look from your listeners when you are showing a picture of a bison, and you are talking about the Black Hills of South Dakota, then be aware that the audience may not be making the connection that you may be taking for granted. Yes, there are bison in the Black Hills, but the audience may not know this and they may be trying to understand why they are not seeing a picture of the Black Hills. At this point, the audience's attention is waning and your presentation's effect is suffering.

Visual aids should be used during a presentation to achieve specific objectives. Here's how they add effectiveness and professionalism to your presentations:

- Use an eye-grabbing title with an image to get the audience's attention. The opening needs to break the audience's preoccupation and draw them into the talk. The opening visual aid can act as 'feedforward.'
- Select visual aids that guide the group's thinking to pre-determined conclusions. Know the audience so that they do not misinterpret the meaning of the visual aids.
- Emphasize key points, not every sentence.
- The most common problem with visual aids is that communicator's tend to clutter their presentations. Resist trying to put the whole story on the screen. Focus on key words and simple graphics, enough to help people keep track of the theme and supporting ideas.
- Give the audience a reason to listen to you. If a visual aid is complete enough to use as a handout, then use it as a handout.
- Present complex, detailed data in understandable ways.
- Explain new concepts, pictures and diagrams to help clarify the details.

Visual aids should be graphic because most people are able to think in pictures. If using word-only visual aids to reinforce the message, keep the number of words to a minimum. If you have a powerful and flashy visual, give the audience several seconds to absorb it and then continue once the 'awe' factor has worn off. Color is useful and can enhance visual aids, but can also be distracting in certain combinations. Select colors to draw the eye to key points that the speaker is addressing. With today's superb and easy-to-use computer graphics, it is simple to get carried away with too much color and graphics.

Questions to think about with visual aids are:

- Is the aid clear? Is it obvious at a glance what the visual aid is trying to communicate?
- Is it readable? Test the arm's length rule – the image should be as readable at a distance as it is at arm's length. Hence the distance of the farthest part of the audience from the visual aid needs to be anticipated, so that the size of the graphics or fonts are large enough to see as though they were only an arm's length from them.
- Does it communicate a single idea? If there are too many points on a visual aid, then the audience may be looking at the wrong point while you are speaking.
- Is it relevant? Does it make a point that fits in with the presentation? Don't reveal a visual until you're ready and remove it when you're finished with the pertinent section.
- Is it interesting? Does it keep the audience's attention? Graphics should capture attention.
- Is it simple? Is it too cluttered with pictures, graphs, colors, borders and lines? Make the focus be the point at which you wish the audience to look.
- Does it support the content? Remember 'Say dog, see dog.'

15.3 You Don't Always Have Electricity

In today's high-tech world, especially in more-developed countries, we seem to take for granted that we will always electricity and high-tech capabilities wherever we go. Do not be fooled. Despite a standardization of electronic platforms, computers and projection equipment still have major connection problems. Technology always seems to fail when you need it most. Outside of urban centers worldwide, you will often find yourself in an electronic void where whiz-bang toys do not work. Indeed, if you live in an urbanized area, go to your local nature park and do an outdoor talk. Chances are you will have no electricity and only the barest of equipment to convey any visuals. A good communicator will be prepared to go from the most advanced capabilities of electronic joy to no-tech at all. If electricity fails unexpectedly, you may really be tested – which curiously does happen too often in our experience.

15.4 Authentic Items and Models

If you are talking about details of a complex topic, such as a proposed visitor center or a food web, then a model or a representative, authentic item can be an incredible benefit to get your audience to understand your ideas. Nothing quite substitutes for seeing and feeling a real object. And, scale models are almost as good. If a model or real item can be transported easily, then they offer opportunities for an audience to learn the intricacies of the topic that would be hard to convey with words alone.

When using real objects and models, be certain the entire audience can see them. Too often, only the front row of an audience can easily see a displayed object, causing the rest to lose interest. Consider passing the object around the room, if feasible. Still, this may not solve the problem, since the speaker may be discussing another point by the time some audience members receive it.

15.5 Warm Fuzzies

Actors in Hollywood always warn of the dangers of working with animals. They tend to upstage you and become the center of attention. If you are talking about an animal and need to indicate certain features of its anatomy or aspects of its behavior, it is wonderful to use it as a prop. But, as soon as you depart from needing the animal, it should be put out of sight of the audience. If the animal does not go backstage, the audience may just keep watching the animal, which distracts them from your message.

A real story of such animal antics comes from a raptor center in the American Midwest. The speaker had a barn owl that had been blinded in a road accident. While the talk centered on the owl's story, it was wonderful to observe the rescued creature. But, when the speaker progressed to other topics such as duties of the raptor center, the owl remained on the speaker's arm. Nearly everyone was focused on the bird, watching how it kept turning it head in response to noises from outside the room. They had all stopped listening. When the bird finally ejected feces onto the floor, the room erupted in surprised laughter. Yes, it was entertaining, but nobody heard the message about the raptor center. Whether in an interpretative talk at a park or in another setting, beware of the warm-and-fuzzy appeal that animals have on an audience.

Another speaker in Montana was talking about wolves. The speaker had both a large dog as well as a captive-reared wolf for the presentation's first portion. The speaker compared and contrasted the two animals in an entertaining and informative way. When the talk moved to other topics, the speaker had an assistant take both animals out of sight. The audience now had been aided visually by the speaker and kept listening as she moved onto build her theme. As the speaker talked about an experience of meeting wolves in the wild, the audience was not distracted by unpredictable antics of the animals, but instead remained focused on the storytelling.

Note that in both cases, the speaker was an expert with excellent interpretive skills. In the case of the owl, the speaker was upstaged by the bird. The speaker with the wolf, on the other hand, kept complete control of the situation and hence the attention of the audience.

15.6 Flipcharts, Chalkboards and Whiteboards

Common forms of visual aids became familiar to many of us in elementary school. Flipcharts, chalkboards and whiteboards are especially appropriate for interactive sessions where the presenter has to capture input from the group or develop lists generated during interactions with the audience. When using write-on media, avoid talking completely to the board or chart when writing. While writing you should stop and look at the audience to reinforce that you are talking to them. If you have so much to write that you have too much 'dead time' when you are not speaking, switch to prepared media or use an assistant to scribe. You will want to carry your own supplies of pens and chalk.

15.6.1 *Flipcharts*

The best uses for flipcharts are for (1) making lists, (2) outlining steps in a process, (3) sequencing ideas, (4) drawing simple sketches, and (5) recording group work such as brainstorming.

Advantages and disadvantages to using flipcharts are:

Advantages

- Spontaneous with little preparation needed.
- Easy and inexpensive to use. They require only a stand, large sheets of paper and pens.
- Can be used in normal lighting.
- Used sheets can be easily posted for re-emphasis during ongoing discussions.
- Pages can be saved as a record of the discussion.

Disadvantages

- Turning your back to audience to write is necessary.
- Not good for detailed information.
- Time-consuming.
- Difficult to move and to use where there is wind or no walls.

Tips for Flipcharts

- Use two flipcharts – one for prepared materials and the other for spontaneous comments and lists. Or, use one for questions and the other for responses.
- Keep statements short and simple to minimize writing.
- Write a title on every page. Titles add impact and help organize.
- Colors help. Use colors with the greatest visibility: black, blue and green, in that order. Avoid purple, brown, pink and especially yellow. Red should only be used as an accent, for bullets, underlines, arrows, etc. Keywords may be written in red when everything else is blue or black.
- Two colors in combination on a chart are better than one. Three is OK, if done carefully and with purpose. More than three makes it difficult to pick up points of emphasis. Combinations to avoid because of contrast and poor visibility are red and green, orange and blue, or yellow with any other color except black. Good contrasts are red with black or blue, and green, yellow or blue with black.
- Lettering needs to be consistent and neat when used on a chart, otherwise it can get distracting. Use print, not script. Use upper and lower case. Using grid lined

paper will help keep lettering straight and allow you to line up margins, subheadings and bullets.
- Leave generous amounts of white space to make the chart look cleaner and the writing easier to read. Try to keep a margin of three to four inches on each side. If points consist of a word or two, a seven or eight inch margin may be appropriate. Whatever margin you leave on the left, try to leave a similar margin on the right.

15.6.2 Chalkboards

Most of the same tips for flipcharts apply for chalkboards and whiteboards as well. Some extra differences that will enhance your use of chalkboards are:

- Keep it clean.
- Use colored chalk for emphasis.
- Prepare extensive drawings before session starts, or use handouts or other media.
- Use paper to cover the board until ready.
- Maintain eye contact with the audience. Don't talk to the board.
- Print words neatly and large enough for all in the audience to see. Remember the back of the room needs to see.

15.6.3 Whiteboards

For whiteboards, keep these in mind:

- Use water soluble felt-tip markers for drawings and lettering.
- Use water-soluble hair spray to protect complex drawings.
- Use colored pens for emphasis, remembering the color combinations given for flipcharts.
- Don't stand in front of what you have written.
- Use templates made from cardboard.
- Outline drawings lightly prior to the session.

If you are left-handed, chalkboards and whiteboards present special challenges. As a southpaw, if you are unable to write clearly and cleanly on these media, have someone else do the recording.

15.7 Handouts

Handouts are visual aids that you distribute to an audience so each person has their own copy. You may choose to distribute an outline of your presentation, so that the audience can follow your discussion. Or you may want to give them essential information, so that they do not have to take notes while you are speaking. You may also give the audience a summary of your discussion, especially if there will be several speakers and you want them to have a reminder of your talk. Finally, some speakers use handouts to give the audience supplementary material that can be used for later reference.

When deciding when to distribute handouts, you do not want to disrupt your presentation or distract the audience. The following are the preferred times to distribute handouts:

	Prior	During	After
1. As an outline	Best	Poor	OK
2. Material essential to discussion	Best	OK	Poor
3. As a summary	OK	Poor	Best
4. Supplementary material	Poor	Poor	Best

Take care to not overdo the volume of handouts you give to an audience. This will tend to confuse and distract from your presentation. If you have many sheets, bind them into a booklet and use it as a workbook that complements your presentation. Leave white space on each page so audience members can enter their own thoughts about your presentation. Color code sections where appropriate. This is especially true when the handouts cover complex concepts, contain diagrams, or have multiple pages which go together within a single point.

15.8 Even When You Do Have Electricity

You are in luck. You have electricity. Yet, now the electronic beasts you expected to communicate with have gone on strike. There are always the old standbys. If you even suspect you may not get your PowerPoint or other software option working at some odd, out-of-the-way location, you might want to invest a few minutes into creating backup transparencies while still successfully hooked up to a computer and printer. At the very least, take some blank transparency slides and a batch of colored transparency slide pens (check they are all working before you go). And yes, be prepared to even use a chalkboard or whiteboard if one is still available and they have not yet been thrown out.

15.9 Overhead Projectors and Transparencies

'Overheads' are still the most widely used method of presenting visual aids globally, because of their ease of preparation and versatility. Transparencies can be prepared with a computer and printed on any printer or by duplicating onto a transparency sheet using a photocopier. Be sure you have the right transparency sheets for the appropriate device. For instance, laser printers and photocopiers use transparencies able to take high temperatures, while inkjet printers use transparencies with a special surface able to receive the ink.

There are advantages and disadvantages to transparencies:

- Advantages
- You can face your audience. This benefits all presentations.
- Normal lighting allows the audience to look easily at the presenter and the overheads.
- Flexibility in preparing, editing and revising since they are easy to make.
- Since overheads are close at hand you can answer unanticipated questions which may refer to an overhead from earlier in the presentation.
- They are easily portable from one place to another, usually not taking up more than a single file folder in size and weight.
- Disadvantages
- Burned-out bulbs can be a nuisance. Many projectors have spare bulbs in them, but you need to know how to change the bulb without electrocuting or burning yourself.
- Glare from some overhead lights may cause the overhead to wash out. You can often just switch off the offending light or reposition the projector.

Techniques for using transparencies and overhead projectors are in Box 15.2.

15.10 Slides

With the advent of digital photography now becoming the norm, photographic slides are now fast becoming the dinosaur of the visual aids options. We, however, include them here for completeness and because there are places where the only technology hidden away in a cupboard is a slide projector – this still seems true of many developing nation electronic cupboards in rural settings. The techniques in using slides and their limitations are equally true for overheads and projected Powerpoint slides.

Slides produce quality graphics and visual aids from a camera, special film, and processing into special mounts which are inserted into carousel trays. Trays fit onto a slide projector for presenting. Many professional camera shops can still develop and mount the slides in about an hour. Many computer programs now available can also produce slides of photographic images and computer generated graphics.

A healthy dose of practice is required to present a memorable slide show. Try the following ideas (most apply to PowerPoint also):

- Prepare the audience before the lights go down. Feedforward is particularly important here to get the audience thinking about what they are to be seeing.
- Leave some lights on in the room if possible or use dimmers. This will help the audience take notes if appropriate, or at least stop them from dozing off too readily in the anonymous darkness. It is important for the presenter to speak quite dynamically since many of the usually non-verbal enhancers are subdued in a darker room.

- Use an average of one slide per 15–20 s. If you need to talk about a slide for more than 40 s, think of some way to split the slide into two. If using a sequence of slides as illustrations, pause several seconds per slide so the audience can absorb the visual. Pointing out one or at most two keys points per slide in the sequence will focus the audience on what is essential in the slide.
- Break presentation into segments of about 5–9 slides that cover a particular topic. Too few slides and the audience may start to feel confused; too many and they may start to get bored.
- Review key points of the presentation at the end.
- If there are large amounts of information being presented, prepare an accompanying worksheet to review and summarize.

15.11 Video and Audio Clips

Recorded sights and sounds can wildly enhance your presentation when used correctly. Many materials are already on the market that can serve your purpose, and new packages are released all the time. You should be familiar with the media before you present it to the audience, however. Your first time operating multimedia software should not be public.

Consider these questions when incorporating video, audio and other multimedia:

- Is the film or video self-contained? Does its message match yours? While much of a clip may be suitable for your purposes, it does require that you preview and mark just how you will use the footage. If you wish to stop at key points and need to fast-forward or rewind to other parts of the media, consider editing only the segments you need that fit with your presentation.
- Preset the equipment for volume, color and focus. Set the clip where you want to start.
- For TV or monitor viewing, be certain all members of the audience can easily see the screen. A rule is that there should be one inch of screen per person in the audience (e.g. a 19″ TV monitor is good for about 19 people gathered near the monitor). If you have more people than a screen can handle, you need a projection system.
- Since you already know the clip, watch the audience to see how they react. Use this quick formative evaluation when you begin speaking to them again.

15.12 Computer-Generated Images and Programs (PowerPoint, Keynote)

Electronic presentations follow the same rules for use of color, font sizes and simplicity of content as any other visual aid. The most widely used presentation software is Microsoft's PowerPoint. Computers have made it easier to go 'overboard' when faced with endless options, though. The advent of computerized graphics has made presentations much more impressive. Still, it should be remembered that not all locations will have facilities to use computerized graphics. Even if you do a computer-based presentation, remember the following admonition 'If it can go wrong, it will.'

The following experiences seem to show technology conspiring to disrupt presentations:

- If you don't take your own computer and projection system (carrying along two bulky cases, although modern equipment is getting much more compact) your file will not load onto the host systems. At a 2009 international conference, all PowerPoint files had to be loaded on to a central system at the conference center through a bank of dedicated upload-only computers. The system refused almost every file the first time someone would try to upload. It was disheartening to see people from all over the world struggling to get their PowerPoints accepted. The lesson: go to the next bullet below.
- Compatibility is a myth. Every piece of electronic equipment may act as though it is a singular prototype – even with the same manufacturer's name on them and the same program specifications. Many people found that many of their graphic images, even when saved as instructed, did not come up on the presentation. The two major computer platforms, Mac vs. PC, sometimes talk to each other just fine, sometimes with trouble, and sometimes incomprehensibly to each other. Older versions of Mac PowerPoint are incomprehensible to a PC, and moving from PC to a Mac seems to depend on the mood of the Mac that day.
- Telephone line link-ups work perfectly until two minutes into the presentation, when they unexpectedly disconnect.
- Brand-new equipment is always suspect. Consider it guilty until proved useful and reliable.
- The host system has a newer version of the program than you do, but it identifies your file and loads it. However, all program updates are so changed from the previous versions that you need a day just to work out the changes. The crucial icons you use are now buried in another menu.
- The host system has a lower version than you do and thinks your file is just a mass of funny faces, geometric shapes, and odd squiggles.
- You have a Mac and they have a PC that wants to erase and reformat your system! Or, vice versa.
- You keep your flash drive clipped to your keychain. You did not bring your keychain with you since it causes troubles in passing through metal detectors.
- Projection equipment light bulbs work for years until you begin your presentation.
- Your cables and adaptors are the wrong size and configuration for the host equipment.

Even if you anticipate everything you can, you can still fall prey to electronic gremlins. One speaker from the United States went to China to do a presentation. After talking by phone and email to the hosts, who were proud of their new electronic theater set-up, the speaker was assured that all that the speaker had to bring was a file disk with the presentation saved on it. The speaker saved the file in four different formats and versions of the program. There was not just one disk copy, but three, and they were kept in two different pieces of luggage and one on-person. The speaker even memorized the layout of the main program icons in case the computer was using Chinese icons. When the day for the talk arrived, one of the versions of the file loaded fine on the computer in the Chinese theater projection booth. There was even a remote 'mouse' for controlling the computer from the stage. The computer was using Chinese, but the speaker was able to 'click' the correct icons to control the file during the presentation. Then, when an obvious error message suddenly came up, the speaker had to guess what was written and just click out the message box. The message box disappeared and so did the whole file, and the computer shut down! Fortunately the speaker here had backed up all the presentations onto overhead transparencies and was even ready to talk just using a white board that was in the room.

Whenever you are using electronic equipment, be prepared for all contingencies. One of the first major decisions is whether to take your own equipment or to rely on unknown equipment at the presentation location. Calling ahead and getting exact details of the equipment is the best you can do for yourself to avoid unforseen problems. Always remember to:

- Have spare bulbs for equipment
- Have extension cords
- Have three-prong adapters or the selection for whatever country you are in
- Turn the computer on, turn *down* lights (avoid dark). When finished, turn *on* lights then turn off the computer
- Practice so there is no 'dead time'
- Have a contingency plan

15.13 Conclusion

Visual aids are a valuable method to augment your presentation. As long as you and your message remain the central components of your presentation, appeal and comprehension can be added to by visual aids. Computer-based technology is likely to become easier to use and more widely available, and is now as commonplace as overhead projectors in many places. Whatever the future trends of graphic enhancements, remember that is all they are – enhancements to your message. Things will go wrong and equipment and power will fail, often at the most inappropriate times. Do not be discouraged. It happens to everyone, and your audiences will be sympathetic. What will set you apart as a professional is how you handle such a situation. Be prepared. If everything electronic should fail, then plan how you might use whatever other resources are available. The simpler graphic aids such as flipcharts, chalkboards and whiteboards do not need power and allow you to build your presentation from scratch. Consider them all and become competent in their use.

15.14 Case Study: Using Visual Aids. 'Thirst' Presentation by Jeff Brenman of Apollo Ideas

This beautiful and potent PowerPoint won the 2008 World's Best Presentation Contest, held by the specialized social network SlideShare.net. It marked the second award in a row for designer Jeff Brenman, of Raleigh/Durham, North Carolina, USA, and his firm, Apollo Ideas. Brenman grew up with a passion for storytelling and design, so he crafted a career around empowering others to communicate their ideas in exceptional ways.

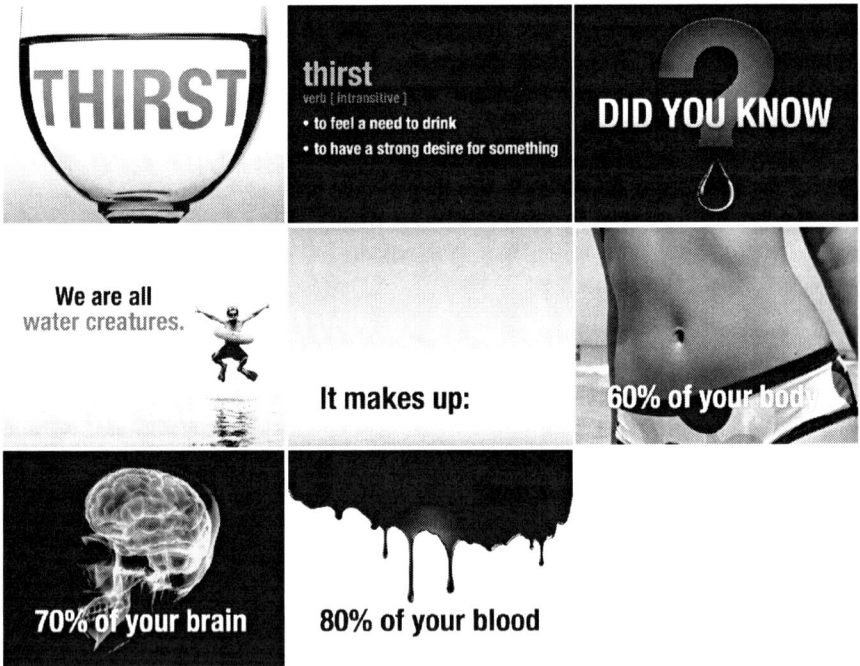

Credit: 'Thirst' PowerPoint presentation, slides 1 and 3–9, at 63 total slides, by Jeff Brenman of Apollo Ideas (used with permission)

Box 15.1 Do's and Don'ts' of Powerpoint

- **Organize and Prepare** – Before you even think of working with powerpoint, organize your thoughts on paper, or through your word processing – remember the message is first, not the images you would like to use. Choose your data carefully so it complements what you wish to say – not vice-versa. Don't work on the visual part of the presentation before having inserted all the necessary text.
- **KISS (Keep It Simple Stupid)** – Just because you have a million buttons doesn't mean you have to push or use them. All those colors and wonderful

(continued)

Box 15.1 (continued)

animations can be entertaining – so much so they can steal the show from you, or as usually happens, the audience is so 'flashed out' they are no longer hearing the message.
- **Visuals** – Focus all the visuals to enhance keywords and relevant data. We are visual creatures so do use visuals – a presentation without them can be as tedious as someone just reading the slides (see below).
- **Ensure the Visual Aids Support the Message** – Remember, Say Dog, See Dog. Powerpoint is merely a tool, not the focus of your message.
- **Minimize the Numbers of Slides** – Don't overwhelm the audience with too many figures and numbers on the screen or too much information. Having said that, don't have just one slide up for several minutes. Many very simple slides is better than a few very complex ones. If you are talking and the slide is not relevant, blank the screen through use of a 'black slide' or with a powerpoint 'clicker' that blanks the screen. The slide should be up long enough for the audience to digest it, in conjunction with what you are saying. Slides that move as fast as an international table-tennis ball in a championship match serve no purpose except to frustrate your audience.
- **Consistency** – Use similar backgrounds, colors, fonts, and slide transitions.
- **Don't Read the PowerPoint** – If everyone in the audience is illiterate, you would want to use pictures anyway. PowerPoint works best with spoken remarks that augment and discuss, rather than having the speaker read what's on the screen. The audience came to see you, not the back of your head.
- **Time your emphasizes** – A well-planned PowerPoint brings up a new slide, gives the audience a chance to read and digest it, then follows up with remarks that broaden and amplify what's on the screen.
- **Give It a Rest** – Don't be shy about letting the screen occasionally go blank.. Not only can that give your audience a visual break, but also allows for you to focus attention on more discussion or a question and answer session to engage your audience.
- **Use Vibrant Colors** – You should aim for a striking contrast between words, graphics and the background. Be aware of those that may be colorblind.
- **Import Other Images and Graphics** – With the internet, there is a wealth of images you have access to (e.g. http://Images.google.com) that can seem almost tailored for what you wish to convey. Some sites (e.g. You Tube) even have video clips you can download to sue on your presentation. However, be sure you have copyright to use these media and even if public domain, give a citation on the source where appropriate.
- **Distribute Handouts at the End** – Not during the presentation. See our discussion about handouts above.
- **Edit, Edit, Edit** – Once you're finished drafting your PowerPoint slides, assume you are one of the audience and focus your talk on them – not just what you want to tell them.

Box 15.2 Techniques for Successful Presentations with Overhead Slides (Applicable to Powerpoint)

- In many meeting places screens have been set in place, but often you will find that you have to still use the good old free-standing screens. Place screen to the side or front corner of room if possible and face the audience. This gives you more flexibility to manage the room and help the audience see the overheads. It also helps create a more relaxed atmosphere if the presenter wishes to move away from the projector occasionally. Stand beside the projector facing the audience so that all slides can be read by you looking at the projector and the audience looking at the screen.
- Use pointer, or if using overheads use pencil as pointer (not your finger) over the actual overhead while it is on the projector. Be aware if you are blocking the screen to any of the audience.
- Use 'disclosure technique' to keep audience focused – this is easily done via animation on Powerpoint. Don't reveal all the information at once. On an overhead, place a sheet of paper over the overhead and move it down as you speak revealing just the section you are presenting at that moment. Cover the screen before removing one overhead and replacing it with another to prevent distracting light glares. You might use a black slide in Powerpoint, although some of the new Powerpoint slide changer 'clickers' have a black slide function on them.
- Speak with more volume than you normally use. The listener's attention is divided, and more volume is needed to hold their attention.
- Use color to add life to overhead transparencies. When using colored transparency film, be careful to check that the overhead is readable on a projector. Some colors can wash out and do not project well. Use overheads creatively, e.g., multiple overlays, colored highlights, changing images, to maintain audience interest.
- Check legibility of visuals from the back of the room. Don't use only capital letters or all italics. Avoid transparency sheets that are dirty. Try to avoid distracting light leaks.
- Don't overload your slides. A reproduction of a typewritten page is one of the worst transparencies there is. Think in terms of bullets and single words or short statements.
- The blank space between lines should be 1 1/2 times the letter height.
- Transparencies of forms are useful to demonstrate how to complete the form. You can write on the sheet with water-based pens. But, be careful of doing this on inkjet transparency sheet because it will not wash off. Use another blank water washable film over the inkjet film if you need to reuse the inkjet transparency.
- When not in use, turn the projector light off. Turn it back on when you need it later in the presentation.

References and Further Reading

Brenman J (2008) Thirst. http://www.slideshare.net/jbrenman/thirst. Cited 25 July 2009
Duarte N (2008) Slide:ology: The art and science of creating great presentations. O'Reilly Media, Sebastopol, CA
Gillin P (2008) Secrets of social media marketing: How to use online conversations and customer communities to turbo-charge your business! Linden, Fresno, CA
Hager PJ, Scheiber HJ (1997) Designing & delivering scientific, technical, and managerial presentations. Wiley, New York, NY
Ham SH (1992) Environmental interpretation. North American Press, Golden, CO
Harris RL (2000) Information graphics: A comprehensive illustrated reference. Oxford University Press, New York, NY
Hooper JK (1997) Effective slide presentations: A practical guide to more powerful presentations. Fulcrum, New York, NY
Kearney L, Wilder C (1996) Graphics for presenters: Getting your ideas across. Crisp, Menlo Park, CA
Leech T (2004) How to prepare, stage, & deliver winning presentations, 3rd edn. AMACOM, New York, NY
Meerman D, Wiley S (2008) The new rules of marketing and PR: How to use news releases, blogs, podcasting, viral marketing and online media to reach buyers directly. Wiley, New York, NY
Morrisey GL, Sechrest TL, Warman WB (1997) Loud and clear: How to prepare and deliver effective business and technical presentations, 4th edn. Basic Books, NY
Rose G (2006) Visual methodologies: An introduction to the interpretation of visual methods. Sage, London
Rozakis LE (1999) The complete idiot's guide to public speaking. McMillan, London
Weissman J (2008) Presenting to win: The art of telling your story. FT Press, Upper Saddle River, NJ

Chapter 16
Dealing with the News Media

16.1 Introduction

This chapter is meant to help communicators understand how news media work, and to give advice on interacting with journalists. The aim is reporting of fair and accurate environmental information after the information leaves control of the originating environmental entity. Fruitful interactions with reporters and producers can translate into more and better coverage as part of an environmental campaign.

There are many misconceptions about news media and their roles in society. It is often assumed news media give a complete picture of current events or situations. While news reporters do aspire for completeness, the news process has many limitations that restrict how much news can be reported and how detailed such reporting can be. The dawn of the digital citizen age of journalism has democratized some of the news process. Even so, widest circulating news is produced by professional journalists working within some form of corporation. Environmental communicators find it imperative to practice utilitarian media relations, even if they are skilled at pushing information through digital channels also.

16.2 What Is the News Process?

News media are not possessed by an altruistic desire to publish and broadcast information from environmental groups just to nudge the public toward environmental literacy. News is a business, one where there is much competition for the reader, listener or viewer to use specific sources of information. Because of this fierce competition each news source needs to be able to offer content that attracts and holds a core loyal audience. Most news outlets do not make money selling their information and rely on advertising that accompanies the news. Look at a typical news web site, newspaper or TV news show and notice how much advertising is present.

News media are processors of accounts of current events which are transmitted to a mass public. News is information and news corporations still control much of what and how we know of happenings in the world beyond our immediate observations. News is event-driven, so it must have some novelty to it. Everyday occurrences usually lack this. News worthiness depends to a degree on the cultural background of both the journalists and that of the intended audience.

16.3 Role of the Media

While news media strive to remain fair and accurate, numerous factors interact to influence how people think about and react to information that is presented. Reporting can make a difference by stimulating certain actions over others, by highlighting specific events and problems at the expense of different happenings. Journalists act as 'gatekeepers' of information in that they control and interpret much of what we read and see. News corporations make economic, social and political inferences about information for their customers, the news receivers. Media news coverage is a net, not a blanket. Only big stories get caught. There is only so much space and time in which to cover information in the news outlets. In traditional broadcast news, information is limited to news holes where time is the major constraint on what is transmitted. News media sell information; they do not supply it free of charge. To get people to 'buy' news coverage means using a marketing approach. Each news media outlet has to entice customers to buy their package of news over a competitor's. Consequently, complex issues get explained for an audience often unprepared with technical background knowledge for deeper understanding. Because of this environmental and scientific issues can, through news processing, get:

- Distorted – Information is altered to make it more understandable or appealing to an audience that would not otherwise be interested in it
- Sensationalized – A personal angle may be given to a story that eclipses the science or environment information that prompted the story
- Oversimplified – The science behind the story is too complex for the audience to thoroughly understand, and through simplification the information becomes erroneous and even misleading
- Inaccurate – Due to news reporting constraints, facts presented are incomplete or fail to give adequate context for understanding the reality they purport to describe

16.4 News Reporting Constraints

Reporters aspire to be fair and accurate. The daily grind of reporting places severe constraints on these laudable aspirations, however. Reporters work for a system in desperate need of fresh stories and new perspectives on older stories as soon as possible after they occur. Old news is not 'news' at all and it does not sell. Friedman (1983) stated a classic list of constraints for journalists in reporting timely environmental news:

- **The 'Scoop'** – To report something first, before any of your competitors do, gives a journalist a rush as well as considerable bragging rights. A scoop is hot news for a short time and holds high momentary appeal to the audience. The biggest trouble with scoops is they do not last. As fleeting as they are, scoops remain a coveted commodity for journalists. If an environmental communicator can offer a story that is exclusive and weighty, one may find success in peddling the story as a scoop to a trusted media contact.
- **Short Deadlines** – Rare is time for in-depth research or multiple source checks to confirm information. Most always, deadlines are looming and so confirmations may not be always as reliable as desired.
- **Editorial and Advertiser Pressure** – Biases from public agencies, private experts and reporters may enter into news reports. Since news media rely on advertisers to make money, the news being reported subtly meets certain expectations or else patronage may be withdrawn. Likewise, an editor is responsible to superiors to maintain certain expectations of information coverage.
- **Lack of Consistent Sources/Knowledge** – Complex technical information is the bane of news media. When 'experts' are in conflict over findings, it becomes difficult for reporters to judge which views hold more weight and are more valid. Stories often end up structured as two opposing sides each represented by a single source. Though such structure appears, on the face, to be balanced, the ebb and flow of scientific debate is missed. So, too, any preponderance of the evidence. Continual peer-review is at the heart of the scientific process, where research is critiqued many times over, and facts are never really proved. Most reporters are not trained scientists in any way, shape or form, so they rely on experts to put the information into non-technical vernacular for them. Breaking technical information down so far is something few experts are very good at doing.
- **Crisis Orientation** – High-profile stories, especially of disasters, get milked for all they are worth. Recurring examples of this type of instant sensationalism include wildfires along urban-wilderness interfaces, oil and chemical spills with evacuations, and tropical storms resulting in huge property damage and deaths. News media become engrossed in covering the inherent human dramas and, only rarely, dig deeper to investigate root causes of these problems and any scientifically-vetted solutions to them.

16.5 Accuracy in News vs. Accuracy in Science

A related mismatch between scientists and journalists revolves around conceptions of the term 'accuracy.' Lewenstein (1997) explained, 'For journalists, accuracy means getting "the facts" right, on deadline. For scientists, accuracy is equated with truth, with taking the time to test information against misinterpretation before expressing an opinion.' This harkens to the on-going conundrum of science journalists, noted by Slosson (1922), as a dual between 'comprehensible inaccuracy' and 'incomprehensible accuracy.' A scientist will plod toward a sole authoritative report based on empirical evidence accumulated over as much time as needed to get a clear picture. A journalist will gather interviews from sources with differing views, set them side-by-side so that, in their contrasts and similarities, a snapshot emerges, and will do so at a harried pace.

As diametrical as these approaches may have been at one time, scientists and journalists have developed a symbiotic relationship, with most professionals attuned to their shared purpose of unveiling truths about society and our environment (Peters et al 2008). In a survey of 1,300 epidemiologists and stem cell researchers in the five leading research and development nations (France, Germany, Japan, United Kingdom and United States), Peters et al (2008) found frequent and smooth science–journalism interactions. Scientists noted exposure of their research by journalists helped to inform and educate. Despite lingering worries about being misquoted, working with mass media was 'favorable' and 'pleasant,' research was generally 'well-explained,' and benefits to society were seen as a result.

16.6 Other Limitations to Science and Environmental Reporting

In reporting a regular story, the reporter can ask: Who? What? When? Where? How? Why? In science and environmental reporting, uncertainty in the answers to these six basic questions can create problems. For instance:

- The news story is believed accurate by the public, although potentially it may not be.
- Environmental issues are complex. There is no single problem nor is there a single solution. Solutions often involve trade-offs, too. Yet this complexity is often disregarded to make the story reader-friendly and more understandable.
- Environmental issues often occur because of long-term cumulative effects. The layering of the issue's details makes it difficult to find a central focus on which to anchor the story.
- Since environmental issues are so complex, they often get 'dove-tailed' with other social concerns. By linking them with something simpler and more understandable, there is a dilution of the environmental issue. It gets watered down and the need for further explanation dissipates.
- Environmental reporting is interpretive, since it has complexity and ample uncertainty. This can lead to a reliance on anecdotes to help build explanations.

- Dueling contentions about what is important in a complex environmental story frequently stem from varying positions by sources, leaving the reporter at a loss to choose between them. So, positions get reported as relatively equally weighted options.
- Scientific language and symbols, which may have specific meanings in an environmental/science context, are often used in general ways in news stories that may lead to errors of understanding. For instance, the statement 'vaporized as in an atomic blast' might have been stated by a scientist as 'carbonized,' 'combusted,' or 'rapidly oxidized.'
- Sources which make journalists' work easier are more likely to get coverage. But, this needs to be recognized as a form of subsidy.

16.7 News Releases

News media often become aware of a potential news item, especially those occurring within or through non-governmental organizations and government agencies, by means of news releases. While news releases are useful to let news editors know about your organization and its activities, it is wise to send information that is pertinent to a larger audience. Remember, a news release needs to be newsy, have a 'grabber' of interest, and be focused on the audience of the specific news outlet you send it to. Reported news is written in a highly structured format (known as 'the inverted pyramid', see Fig. 16.1).

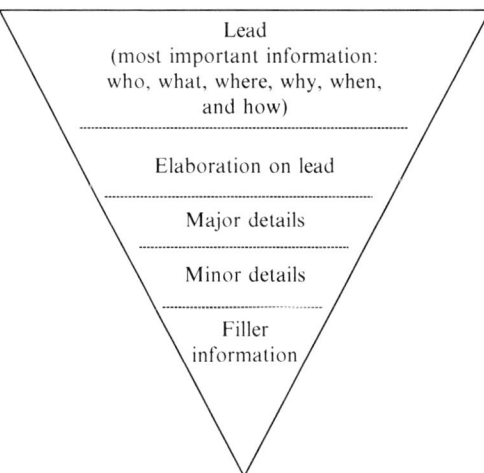

Fig. 16.1 Inverted pyramid of news writing

Inverted pyramid format is highly structured and lets an editor easily chop the length of an article to fit the space available. This respects the always-present deadline crunch by allowing for easy editing without having to search for essential information or doing a major rewrite of the material. It is common newsroom practice to simply delete sentences or even whole paragraphs from the bottom upwards. So, when preparing news releases, essential information must be at the top of the article. Occasionally, an editor may ask for more detail if a story is interesting and space is available, so that the pyramid may be extended with filler information without having to modify the main story already written. Nevertheless, the whole piece is written in as few sentences as possible. Ideally, a news release is shorter than one page.

16.8   News Media Options

When preparing editorial material for submission to news media, an environmental communicator works to combine their organization's campaign with the needs of the targeted outlet's audience. Consider these different outlets:

- **Newsletters** – Organizations generate their own newsletters and so have full control over content. Often, newsletters are distributed to news media as well as organizational constituents.
- **Community Newspapers** – These usually come out weekly and are provided to each household in a neighborhood. Primary information deals with the neighborhood and surrounding communities. They rely heavily on input from their home neighborhoods.
- **Weekly Newspapers** – These serve small, rural markets, filling the gap between media from the closest urban area and local word-of-mouth. They tend to report mainly local information.
- **Daily Newspapers** – The traditional standard-bearer of news, they are published in specific locales and usually circulated in the morning (although a few evening papers still exist). Newspapers come home-delivered, purchased from curbside dispensers, or brought at newsstands. Major papers have regional and national distribution from a central city with branch offices in cities across their circulation area, to handle local issues.
- **Wire Services** – Wire services do not actually distribute stories to mass audiences, they supply them to other outlets that do. Many local news outlets cannot afford to have reporters cover national and international events, and so rely on wire services. News leads in the wire services are constantly and feverishly updated. This feed is known as 'the ticker' in news rooms.
- **Magazines** – Magazines tend to serve narrower audiences and to feature content heavier on analysis and synthesis. Typically, their time horizons are longer than newspapers and broadcast media.

- **Broadcast News Media** – Broadcasters operate on the shortest time horizons of all the traditional media. News releases are best faxed or emailed to a news director, and paired with a phone call alert. If a news release gains interest, then scheduling becomes the biggest hurdle. Be prepared to drop everything at a moment's notice.

Each news media option has a mirror on-line version. The digital frontier actively seeks input from citizen journalists.

16.9 Scientists/Engineers and the News

Scientists and engineers who seek to place items in the news are wise to understand there is a big difference between what they want to say about their research and what journalists will actually report! Reporters write to inform, not educate. Their editors, as gatekeepers, do the selecting and the selling of the stories. Reporters in competition with other reporters have an allegiance to the truth, and are looking for answers that will address the 'greater good' for society.

The reality of science reporting means:

- Reporters are concerned with the application and social relevance of science, not implications to any particular field of science. Hence, scientists and engineers will be more successful in having their work reported if they interact with news media on these terms. Scientists need news media in order to promote science outside the science community.
- Scientists who reviewed science media stories generally concluded that one-third of stories had inaccuracies, omissions or were too vague. Reporters retorted that this one-third represented information of little value to a lay-reader. Scientists do not understand audiences and especially the background limitations of the 95% of the public who are not scientists, critics say. Journalists understand broad audiences more than scientists.
- Scientists need to relate research findings with some logical topic of public interest. For example, an article about slime mold flagella might be related to sperm motility.
- The uncertainty of science is not realized. Scientific statements are seen as definitive.

For scientists to connect better with reporters, they might:

- Wait for publication, or at least acceptance, of findings in the scientific literature, to retain credibility. The classic case of cold fusion in a test tube emphasizes this point. Two scientists, with their home institutions encouragement, contacted news media and held a press conference about the findings of their research in which they purported to have created cold fusion. The significance was simple, safe and endless cheap energy – science fiction turned reality. Unfortunately, the scientists had not published their results in a peer-reviewed journal yet. When other scientists tried to reproduce the results, they could not. The whole experiment was nothing more than a sham (Platt 1998). Yet, the public had been alerted about something that was not real and, in the public's eyes, credibility of all scientists suffered.

- Be prepared to address what is important from the reporter's perspective.
- What are the ESSENTIAL findings? State facts succinctly; limit use of scientific jargon. Remember the audience may not know any of the science involved with the issue. What is the societal importance?
- Forget the gory details of methodology and statistical analysis. If you cannot give a definitive answer, think twice about stating it all to a reporter.
- Analogies are really helpful, especially if some methodology must be given.

To prepare for being interviewed, think through criteria for determining news value:

- Are the research findings new? Why should they interest people beyond the readers of the journal in which they are published? Address the larger significance to society. Because environmental issues are complex, give more than just the basic facts. News must be current or relate to something current.
- Is it unusual enough to be reported? If talking to a local reporter, strive to offer a local hook. Offer a human-interest aspect. Offer an answer the question, 'Why should I care?'
- Is the information source prominent? What are the qualifications of the scientist speaking?
- Keep statements succinct and to the point. If providing written background, provide detail in the correct format (inverted pyramid). If giving an interview only, have talking points to refer to, so information is ready to state clearly and without unnecessary hesitations.

16.10 Conclusion

Working with members of the news media is an increasingly important aspect of environmental communication. News coverage can be extremely effective in transferring messages about resource management issues, but it can also be damaging. Work to avoid crisis communication. An environmental communicator who understands the dynamics of the news process will be better suited for placing favorable stories. With media relations skills, quality and quantity of coverage can be influenced. Positive interactions with news media are most often the result of careful communication planning and a keen understanding of the news process.

16.11 Case Study: Media Relations. Environmental Working Group

A group whose president says 'It's my job to stay pissed off and work even harder to make people think about the kind of world we've built and the world we're leaving our kids,' as Ken Cook states in an annual report (Environmental Working Group 2008), probably does not lack for passion in their work. Environmental Working Group backs up their fire-in-the-belly with focused campaigns and outstanding media relations. One such campaign, started in March 2007, questions the ubiquitous use of plastics chemical BPA (bisphenol A) in places such as baby bottles and linings of baby formula cans. That BPA mimics hormones and may be linked to pediatric health problems was trumpeted. Through reports, web sites and lots of media contacts, the campaign caught on and coverage compounded. Industry groups were concerned enough to fund counter-campaigns. Public attention has so far lead to several outcomes. A raft of new scientific studies on health effects of the substance have begun, since the existing literature is murky. Retailers, including giants Wal-Mart and Toy R Us, felt pressured enough to stop selling BPA-containing products. Policy responses include bans in Canada and the U.S. states of Minnesota and Connecticut. *Photo: Courtesy of Environmental Working Group.*

References and Further Reading

Brooks BS, Kennedy G, Moen DR (2007) News reporting and writing, 9th edn. St. Martin's, Bedford
Environmental Working Group (2008) Annual report 2007. Environmental Working Group, Washington
Friedman SM (1983) Environmental reporting: Problem child of the media. Environment 25(10):24–29
Gause V (2001) A beginner's guide to media communications, 4th edn. McGraw-Hill, Glencoe
Jones C (2004) Winning with the news media: A self-defense manual when you're the story, 8th edn. Winning News Media, Anna Maria, FL
Lanson J, Stephens M (2007) Writing and reporting the news, 3rd edn. Oxford University Press, New York, NY
Lewenstein B (1997) International perspective on science communication ethics. Frontiers: Int J Study Abroad 3(2):187–196
Mencher M (2007) News reporting and writing, 11th edn. McGraw-Hill, New York, NY
Peters HP, Brossard D, de Cheveigne S, Dunwoody S, Kallfass M, Miller S, Tsuchida S (2008) Science communication: interactions with the mass media. Science 321(5886):204–205
Platt C (1998) What if cold fusion is real? Wired, retrieved 6.16.2009. http://www.wired.com/wired/archive/6.11/coldfusion.html
Rich C (2009) Writing and reporting news: A coaching method, 9th edn. Wadsworth, Bedford
Slosson EE (1922) Science from the side-lines. Century Ill Mag 107:471–476
West B, Greenberg M (2003) Reporter's environmental handbook, 3rd edn. Rutgers University Press, New Brunswick, NJ

Chapter 17
Managing Conflict

17.1 Introduction

Human history is full of notable conflicts over natural resources management and environmental issues from all corners of the globe. Indeed, in an article about ecosystem management, the Ecological Society of America pointed out these are not conflicts between humans and nature, but conflicts between competing human needs for natural resources (Christensen et al 1996). This chapter is not meant to be an in-depth analysis of conflict management, but more a guide to help a communicator understand their role in conflict situations. We have chosen the term 'conflict management,' rather than 'conflict resolution,' because many situations arise in environmental issues are not resolvable. Natural resource professionals are best suited to finding ways to make conflicts as manageable as possible, rather than letting these conflicts debilitate organizations and communities.

17.2 Values of the Environment

Conflicts are based in differing values among disputing parties. Five bases of human values can be attributed to ecosystems (Mullins and Watson, 1996):

- Economic output – making money from natural resources
- Ecological services – meeting of human needs such as clean water and climate regulation
- Aesthetics and spirituality – sources of beauty and inspiration
- Ethics – a moral obligation to protect the natural environment
- Education – a place to conduct science and other forms of learning

Nearly all people hold values related to all five bases, but individuals tend to prioritize them in vastly different ways. People are reluctant to risk value(s) they hold most dear. Thus, a rancher may value the educational aspects of a rangeland, but seeks first to protect his economic value for livestock grazing. By contrast, a recreational hiker may be unwilling to compromise on the aesthetic/spiritual value of a wilderness area, even though she supports the idea of scientific research in the area. At the root of conflicts are differing values.

In cases where disputants in a conflict can each achieve protection of the values they prioritize highest, we call the solution 'win–win.' In contrast, when one side of a dispute achieves their goal, while the other side is unable to obtain a satisfactory solution, we have a 'win–lose' outcome. All too often, environmental conflicts are expensive, drawn-out situations in which all involved fail to achieve an acceptable outcome. These are called 'lose–lose.' Skillful conflict management is essential to achieving win–win outcomes to natural resource debates.

17.3   Reasons for Conflict

Conflict can be argued to be inherent in human societies, and it is not necessarily detrimental. For example, conflict prevents stagnation, stimulates creativity, allows disputes to be aired, provides a forum for testing ideas, creates cohesion in a group, and can be a source of renewed energy. Negative effects of conflict, however, are more often considered: resources wasted in competition, creation of misconceptions and biases, decreased communication, blurring of issues, magnification of differences, escalation of conflict, and hardening of positions (Carpenter and Kennedy 1988). The prime strategy of communicators involved in conflicts is to take advantage of positive aspects of conflict, while minimizing dysfunctional effects.

Conflict can occur at any scale. It can be found among individuals, members of a group, or different groups, cultures and nationalities. Indeed, in complex environmental issues, all of these types of conflicts may occur simultaneously. Different levels of conflict are observable in most environmental disputes. While war metaphors are often used in describing conflicts, actual tactics used in a real issue may be more psychological than physical. But, such abuse and disregard can be every bit as damaging. As issues conflagrate past the contention stage, the communicator should judge if the conflict has already reached a higher level with some or all of the protagonists (emphasized by hurt, disgust, agitated anger and possibly hatred). If it has escalated in the minds of some of the disputants, then an expert may be the only one to help bring the issue back to a manageable level.

17.4   Anatomy of Conflict

As disagreements over environmental values arise, a range of behaviors may be expressed by disputants. One may choose to avoid the conflict for the time being, delaying it until a later time when there are likely to be fewer options for mutually acceptable resolution. Disputants may choose to collaborate to seek a true win–win solution. Or, more often, they may enter into a competition, seeking to win the conflict regardless of the impact on the other party. Sometimes, disputants become aggressive, seeking not only to win the conflict, but to inflict political, psychological or even physical damage on the other party (Fig. 17.1).

17.4 Anatomy of Conflict

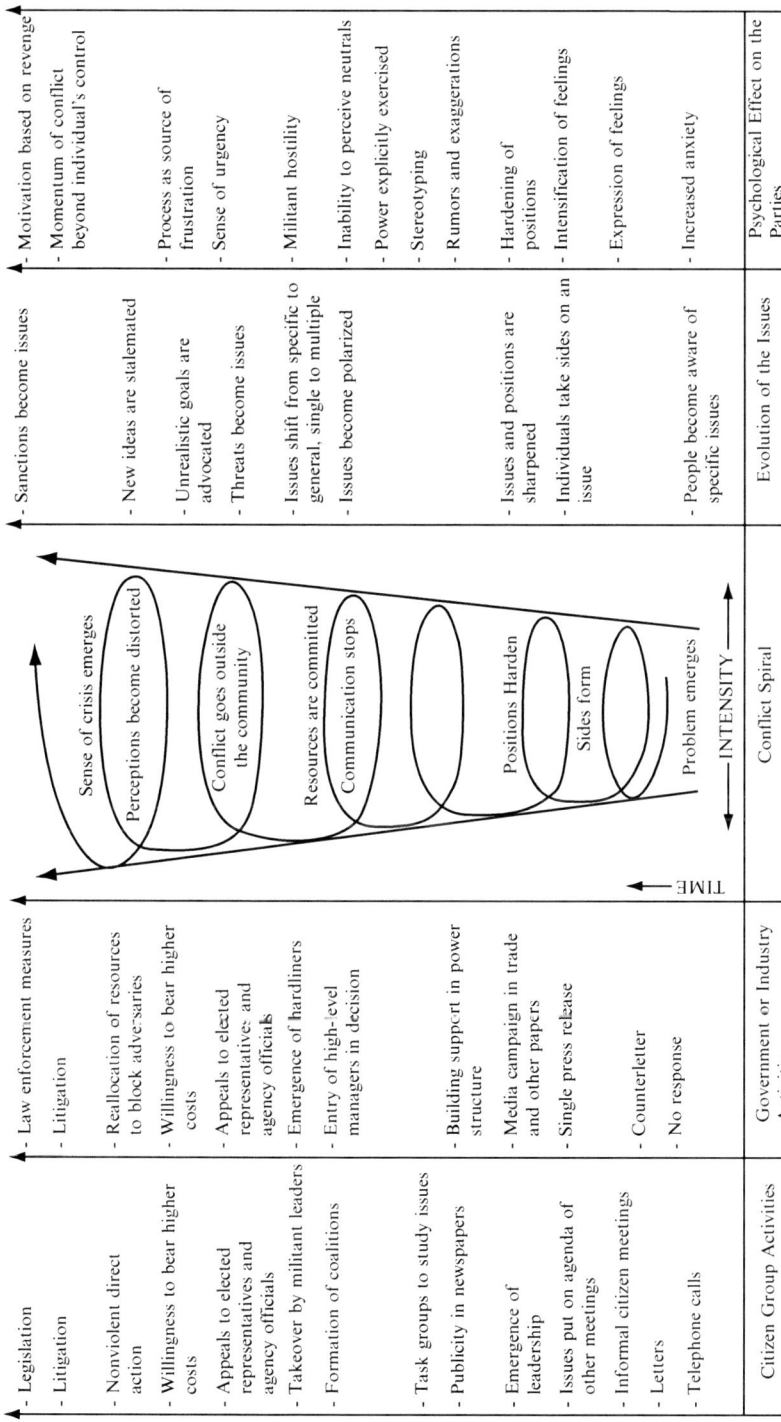

Fig. 17.1 Spiral of Unmanaged Conflict. Carpenter and Kennedy (2001) Managing public disputes. Printed with Permissions US@Wiley.com

Unmanaged conflict develops in fairly predictable ways, as illustrated by the 'Spiral of unmanaged conflict' figure (Lewicki et al. 2005). A spiral depicts the swirling aspects of an escalating conflict. Notice how the two separate groups in the left columns treat the conflict situation as it develops. The right two columns emphasize how the parties begin to take on aspects that were not part of the original problem. Note also in the last column how the mindset changes quite drastically from one of concern to one of a power struggle. Eventually aggression against opponents, and the elevated conflict level themselves, become the issue that seems to consume participants' energies. The cycle becomes vicious. At the top of the spiral, winning may turn into the ultimate goal, even more important than resolving the underlying issue. It should not be construed that the spiral indicates a crescendo with no resolution. Rather it emphasizes that, unless expedient action is taken to understand and manage the conflict, the conflict can grow beyond control.

The spiral suggests one way to constructively manage conflict. Disputants who emphasize shared interests, as opposed to positions, are more likely to find a mutually acceptable solution to their conflict. This is best achieved through principled negotiation. If a dispute can be structured as win-win, it uses a constructive problem-solving approach. To approach successful resolution:

- Understand and act on interests, and not positions, of parties involved.
- Separate people and their feelings from the problem.
- Develop options for mutual gain by all parties involved.
- Insist on fair and accurate criteria (not emotional-subjective ideas) for selecting solutions.

Once positions of where people stand on a situation are defined, it is much more likely that either one or both parties will have to accept a less-than-optimal outcome. Communicators can work with disputants to find acceptability among alternatives for resolving conflicting positions.

17.5 Resolving Disputes

Once toward the top of the spiral of conflict, disputants will be unable to find a mutually acceptable solution to their conflict. When a conflict tops out, there will be limited, mostly unattractive options for ending the conflict. At other spots along the spiral, more options will be available and those options will be less likely to inflict heavy losses on at least one side.

Major options for conflict management are:

a. **Negotiation** – Parties to the conflict work out a resolution between them. They find common ground and collaborate on a consensual resolution. This method negotiates something reasonable through a formal mechanism and can result in a win-win situation. It may require a peer mediator or process manager to keep negotiations on track, because negotiations require process skills and a presumption that the each party's values are legitimate. Accepting values of both (or more) sides in a negotiation is possible if the existing value frames overlap (Lewicki et al., 2005). Compromise solutions are developed that require both parties to give up something.

b. **Third Parties** – Ideally, get an uninvolved third party. All people with a stake in the issue must agree beforehand on criteria for choosing a mediator or arbitrator.

  i. **Mediation** – Sometimes, a mediator facilitates a face-to-face dialogue between the parties. Other times, a mediator is an intermediary, travelling (shuttling) between the parties. A mediator serves as the chief of the process to reach a solution.

  ii. **Arbitration** – An arbitrator is a third party who reviews all the evidence and then makes a decision. The decision may be binding (legal) or non-binding (voluntary acceptance). Details of the arbitration must be agreed upon before it begins. Legal contracts may need to be drafted beforehand to ensure compliance is followed in event of a decision that is not palatable to one of the parties.

c. **Legal Remedies** – Resolutions are determined by officials at a level of government with jurisdiction over the parties. This may be a court, legislature or executive agency.

  i. **Litigation** – The issue enters the legal system. Court systems are monetarily expensive and prone to extensive delays, leaving many options for counter-suits. Having 'standing to sue' equals being qualified to be heard by the court by virtue of being materially affected by the issue. There is also a lack of environmental expertise by many judges, who rule on procedural details and not the content of an issue. Litigation is nearly always an adversarial situation resulting in a win–lose outcome.

  ii. **Rule-Making** – This form of solution involves lobbying for political action in the form of codification, such as rules, laws or regulations. Government agent involvement by its nature is expensive and requires lobbying skills. Access to legislators, for instance, is severely limited. Like the court systems, rule-making can be prone to delays. Assuming proposed bills make it to a legislative floor, there can still be deadlock on controversial issues. Rule-making actions can be pursued at local, state or federal level. At the federal level, local specifics may not be understood and the voices of minority groups may not be heard.

What should be obvious is the increasing amount of time and resources that are invested in the management process to resolve an issue that has circled up the spiral of conflict. The sooner any conflict can be managed and resolved, the less money and other resources expended, and the more likely a suitable win–win result will occur.

17.6 Communicating About Conflict

Carpenter and Kennedy (2001) offered the following guidelines for resolving conflicts. These are appropriate for communicators to consider in planning messages for disputants. Messages are powerful in the charged atmosphere of a conflict. If management toward a win–win is desired, a communicator is obligated to not further enflame the dispute. Messages will need to be suitably sensitive, to state a position without pandering to negative attacks.

- Create a supportive climate in which all parties feel as though their input is acceptable.
- Use descriptive speech and thinking. Resist immediate evaluation. Language is the tool we use to resolve conflict, although its misuse can create or elevate conflict.
- Be flexible and spontaneous. Have a problem-solving attitude, not a combative one.
- Have empathy, not sympathy. Truly try to understand what are other points of view and positions. Understand the problem, in all of its dimensions.
- Demonstrate fairness. Do not be hegemonic.
- Realize that conflicts are not just about substance, but also about procedure and relationships.
- Plan a management strategy, then follow it.
- Build positive working relationships between the disputants (and facilitators, if applicable).
- Begin with a constructive definition of the problem.
- All parties should help design the management process and solution.
- Good solutions are based on interests, not positions.
- Have a problem-solving attitude. Think through what may go wrong. Be prepared.
- Above all, do no harm. You may fail the first time, or have to work with these parties again. You want to garner their respect even if you disagree on principles.

17.7 Conflict Happens

Recognize conflict is inevitable, and is best managed at its outset. Early management prevents escalation beyond control. Points to ponder:

- For what reasons do governments exist? They are institutional arrangements related to legal remedies. A government provides services and manages conflict. It is also for authoritative allocation of values (e.g., the U.S. Congress passes laws such as the Endangered Species Act to uphold societal values).
- Disputes occur over how landowners can use their property. We have social institutions to resolve land use conflicts. Typical conflicts are between and among private users, and between and among public and private users. What the public often wants and what private owners want will be different. Finite land resources and a growing population will inevitably lead to differing agendas on the use of land.

- There are conflicts between haves and have-nots. Those that have ample resources tend to not share equitably with those lacking resources. Examples are water, grazing on public lands, and mining.
- Many people see the need to conserve resources and reduce the impact of existing problems now, while others see new technologies as panaceas for all present and future problems. These positions reflect differing value bases and affect how solutions can be achieved.

17.8 Conclusion

Conflict over natural resources and use of the environment is a given. Conflicts are the source of change in societies. Understanding just how conflicts arise can help a communicator work to minimize or resolve a situation before it escalates out of control. The aim is to foster beneficial changes to society. By working to produce win–win situations, where all parties can gain something of valued importance, conflict in itself can be driver of progress. Conflict becomes destructive when it is ignored or when it facilitates hegemonic imposition on others. By involving the parties to a dispute in the management process and understanding their values, conflict management can be a positive process.

17.9 Case Study: Conflict Management in the United States Allagash Wilderness Waterway, Maine

Thoreau wrote about the Maine woods, which ever since have seen controversies on clear-cutting, habitat preservation, hunting and fishing, and recreation. The Allagash Wilderness Waterway (AWW) was established in 1966 by the Maine legislature, as a 92-mile-long string of lakes, ponds, rivers and stream in the state's remote northern woods. There has been continual conflict about management of AWW, which flared in 2002 when American Rivers, an activist group, claimed mismanagement and the granting of too much access was causing irreparable harm.

To resolve the situation quickly, the new Maine Commissioner of Conservation, Patrick McGowan, convened a two-day retreat of 23 major stakeholders to address the issues and create an agreement of resolution. The belief of many of the participants was that a generation of fighting had hardened the opposing positions with little hope of success. Too many previous administrations had already failed to create any consensus on management and access, they said. So, what made this retreat different?

To begin with, all the right stakeholders were present along with a series of experienced, neutral content advisors. The lead mediator interviewed key participants before the retreat to become fully aware of where shared values and discord existed and to gauge which aspects had hope of consensus early, relinquishing 'hot-button' issues until later. All participants agreed to be able to disagree, especially with interpretation of original intent of AWW legislation. Agreed-on ground rules encouraged flexibility instead of hard positioning. Small-group cohesion exercises, linked to the shared values on the AWW, helped reduce the demonizing each position had for the others.

A number of tentative agreements were reached by the end of the first day, with a final package of compromises from all parties at the end of the next. Rituals of closure and debriefing helped consolidate the agreement. No single interest group claimed victory from the retreat. Rather, there emerged a sharing of interests that persevered and molded the final resolution.

Credit: Conflict Management in the United States: Allagash Wilderness Waterway, Marine. Clipart 'Wilderness' Word 2009 Office.

17.10 Case Study: Conflict Management in Mongolia. Pastoralists vs. Miners

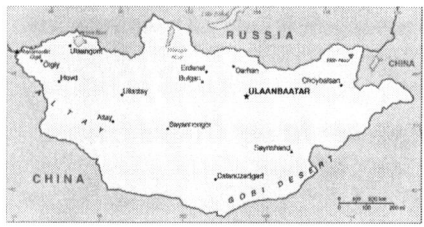

This next case study exemplifies a conflict for which there is no compromise without extreme rethinking of values and ultimately a compromise of values. It raises deep questions: What is the broader cost of modernization and globalization? Who benefits? What cultural assumptions are made about the modern age?

Mongolia has a land area of 1,556,000 km^2, with more than 75% of the land used for grazing, which is the primary industry for the country's economy and source of food for Mongolians. For more than 2,200 years Mongolia had no private land ownership; it was a public commons with locally agreed-to usage of grazing rights. In 2003, however, the national government allowed privatization of some urban areas and some local agricultural land (less than 0.2% of the total land area). Mongolia has substantial deposits of many economically viable minerals, and in the last couple of decades has issued over 500 mining licenses to private companies (often international). Since 1997, more than 147,000 hectares has been degraded through mining activities.

Mining occurs on state-owned land that is separate from public-owned land, a nuanced distinction. But, most lands have been civil-coded as state-owned since 1992. The state has the authority to give land for privatized uses (including subsoil rights as part of contracts), but citizens may not transfer any privatized lands they may have been granted. In 2002, a new law allowed market-based utilization of land, thus repealing a centuries old 'commons law' of open-land usage by its peoples. Herding rights became traditional, not legal.

Mining provides seasonal industrial employment to some pastoralists and helps offset hardship when weather-related factors diminish herd sizes. Unfortunately, this has the effect of eroding nomadic pastoral lifestyle too. Increased human population and subsequent increased numbers of herd animals are also creating problems for nomads. There are now fewer areas on which to graze animals on fresh pasture; overgrazing is now a major concern. While the government struggles with economic transition, a lack of formalized legal rights conflicts with lifestyle practices. As they now exist, nomadic herding and privatized mining are incompatible. The conflict is spiraling upward and attention must be paid to competing values – a socio-cultural history of 'commons law' usage and nomadic lifestyle versus modern mining economics and private sector globalization.

Credit: Conflict Management in Mongolia: Pastoralists vs. Miners. Mongolia map. The World Factbook.

References and Further Reading

Borisoff D, Victor DA (1997) Conflict management: A communication skills approach, 2nd edn. Allyn & Bacon, Boston MA
Budjac C, Barbara A (2006) Conflict management: A practical guide to developing negotiation strategies. Prentice Hall, Upper Saddle River, NJ
Carpenter SL, Kennedy WJD (1988) Managing public disputes. Jossey-Bass, San Francisco, CA
Carpenter SL, Kennedy WJD (2001) Managing public disputes: A practical guide for professionals in government, business and citizen's groups, 2nd edn. Jossy-Bass, San Francisco, CA
Christensen NL, Bartuska AM, Brown JH, Carpenter SD, Antonio C, Francis R, Franklin JF, Machahon JA, Noss RF, Parson DJ, Peterson CH, Turner MG, Woodmansee RG (1996) The report of the Ecological Society of America on the scientific basis for ecosystem management. Ecol Appl 6(3):665–691
Costantino CA, Merchant CS (1995) Designing conflict management systems: A guide to creating productive and healthy organizations. Jossey-Bass, San Francisco, CA
Crowfoot JE, Wondolleck JM (1990) Environmental disputes: Community involvement in conflict resolution. Island Press, Washington, DC
Fisher R, Ury W, Patton B (eds) (1991) Getting to yes: Negotiating agreement without giving in. Penguin, New York, NY
Furlong GT (2005) The conflict resolution toolbox: Models and maps for analyzing, diagnosing, and resolving conflict. Wiley, Mississauga, ON
Greenwood M (2008) How to mediate like a pro: 42 rules for mediating disputes. iUniverse
Hoover J, DiSilvestro RP (2008) The art of constructive confrontation: How to achieve more accountability with less conflict. Wiley, New York, NY
Jones TS, Brinkert R (2007) Conflict coaching: Conflict management strategies and skills for the individual. Sage, Thousand Oaks, CA
Kriesberg L (2006) Constructive conflicts: From escalation to resolution, 3rd edn. Rowman & Littlefield, Lanham, MD
Lewicki R, Saunders D, Barry B (2005) Negotiation, 5th edn. McGraw-Hill/Irwin
Roy L, Saunders D, Barry B (2005) Negotiation, 5th edn. McGraw-Hill/Irwin, New York, NY
Mullins GW, Watson MD (1996) Developing public education packages: A U.S. National Park Service perspective. In: Szaro R, Johnson DW (eds) Managed landscapes: Theory and practice. Oxford Press, New York, NY, pp 593–604

Scott GG (2007) Work with me!: Resolving everyday conflict in your organization. ASJA Press, New York, NY

Sherman R (2004) Get them to see it your way, right away: How to persuade anyone of anything. McGraw-Hill, New York, NY

Weeks D (1994) The eight essential steps to conflict resolution: Preserving relationships at work, at home, and in the community. Putnam, New York, NY

Chapter 18
Communicating About Risk

18.1 Introduction

Ask a scientist what 'risk' means and, more likely than not, they will stress statistics about the likelihood of something nasty happening. But, as communications scholars and social scientists have examined the interface of technology and society more closely during the last 30 years, they have reached a general consensus that non-scientists view risk in an almost antithetically different way from those in the scientific community. Social scientist Peter Sandman (1991) has said, 'The things that kill people and the things that scare people are diametrically opposed.'

For scientists and engineers, and those who communicate about their work, the take-home lesson is this: citing the probability of something negative happening is never enough. If you want people who do not share your scientific viewpoint to accept your information, you will have to do more than spout statistics.

Scientists and engineers who deal in risk are usually driven by numbers and hard data obtained from tests and experiments. While these may be quantitatively valid for setting exposure levels and probabilities of accidents, such scientific observations do not deal with the broader public's hard-to-measure psychological reactions to risks. The non-probabilistic aspects of risk are the focus of this chapter.

When communicating about risk, public trust and the communicator's credibility are the most important factors. Content has been shown to be secondary to these intangibles. A convenient way of looking at the scientific and non-scientific parts of risk is offered by the following equation (Hance et al. 1990; Sandman 1991):

$$\text{Risk} = \text{Hazard} + \text{Outrage}$$

where
$$\text{Hazard} = \text{Probability} \times \text{Consequence}$$
and
$$\text{Outrage} = \text{a cultural reaction to risk}$$

Hazard is the scientific determination of how harmful a particular risk has been measured to be. It is the likelihood of a problem arising from a specific situation, and how problems will manifest themselves should a problem occur. **Probability** is the statistical likelihood that a problem may arise. **Consequence** is the predicted outcome should the problem become real. And, **Outrage** is the perception (real or imagined) of a problem.

18.2 What Is Hazard?

When experts determine risk, the process takes place within the scientific disciplines of epidemiology and risk analysis. Epidemiology is a medical science examining the incidence and distribution of human diseases. Risk analysis deals more broadly with risk from a variety of methodological perspectives. The Society for Risk Analysis (2009), a worldwide organization devoted to the study of risk, defines 'risk analysis' as:

> a detailed examination including risk assessment, risk evaluation, and risk management alternatives, performed to understand the nature of unwanted, negative consequences to human life, health, property, or the environment; an analytical process to provide information regarding undesirable events; the process of quantification of the probabilities and expected consequences for identified risks.

Risk is measured through analogy and mathematical extrapolation from research studies often using models (based on animals, plants or computer programs), and then by indirect comparison to human populations. One of the biggest problems with risk communication is not the quantification of hazards, but explaining the judgments made by risk analyzers as to what is hazardous and unhealthy and what is not.

Hazard is determined by risk assessment, wherein hazards are quantified. In assessment, it is determined how toxic and dangerous something might be. Since most testing does not involve humans, some important information is nearly always missing. With somewhere between 500 and 1,000 new chemicals being introduced to the global marketplace per year, this situation is not going to change. There is not enough time, money or personnel to ascertain the toxic potential of even a fraction of these chemicals. You may be surprised to find out U.S. federal law does not require chemicals to have any risk assessment before they are used.

Risk management is a firm attempt to control hazards in a positive manner. It uses risk communication and quality management procedures to actively control potential exposure to hazards. But, when communicators deal with risk, they are usually dealing with audiences who have perceptions of risk far different than those of the experts who quantify the hazards. This component is outrage.

18.3 Outrage

Citizens and activist groups do not like hazard numbers any more than industry and government officials like outrage statements not based on data produced via the scientific method. So, there is built-in tension in any discourse about risky things.

What a risk communicator has to keep in mind is that perceptions are tantamount to infallible truth to the people holding the perceptions. In one's mind, perceptions are reality. We all believe what we want to believe, but base our views of the truth on different sources of facts. Not everyone finds science to be the best or only arbiter of truth. Regardless of the scientific information involved, much of the time an audience does not understand this technical information. They may not wish to understand it, either. They may even feel alienated by it. But, communicating about risk may not be as hopeless as those assertions make it seem. There are known factors that play significant roles in the perception of risk.

When experts examine a risk, they focus almost exclusively on mortality and morbidity: how many people die and how many people get sick. But, citizens define risk much more broadly, considering a wide range of 'outrage factors' in determining how risky they consider a situation to be. Most always, outrage is not a misperception of the technical data. Rather, it is how people 'feel' about a potential or actual problem in their minds, beyond data attached to mortality and morbidity. There might not even be a demonstrable problem—no one has died and no one has gotten sick yet. But, people are still not happy about the situation. That some group perceives a problem is reason enough to deal with it as a real problem. Perceptions are real. Scientists and engineers need to understand this and deal with outraged citizens seriously and frankly. They are not delusional. Civil discourse about risk, broadly construed, promotes a better society that deals with the hazards and the outrages of modern living.

18.4 Risk Acceptance

For communicators working on environmental risks with scientists, engineers, activists and affected citizens, tactics to keep outrage low are as much a part of risk management as work to prevent mortality and morbidity. Here are some factors involved in risk acceptance:

- **Choice** – Do affected people have a choice on whether they get exposed to potential risks? Sandman (1991) notes, 'The right to say "No" makes saying "Maybe" a lot easier.' When people are part of the decision-making process they tend to accept risks more readily. This gives them some measure of control over new risks in their lives. If any risk is imposed without consent, expect negative reactions. Similarly, if people are comfortable with an existing risk, it is more acceptable even if the hazard gradually rises for some reason. In cases of familiarity connected to a changing hazard, you might actually want to heighten an audience's concern to help safeguard them. Many federal workplace rules and regulations are mandated specifically with such an intention.
- **Financial Burden** – Risks and benefits should be shared equitably. If a chemical company builds a plant and its neighbors bear the entire risk burden, and yet have no chance of seeing any of the profits, they will be outraged. Other examples might be a property–owner whose real estate value is greatly diminished because

of being next to a stream that receives effluent from an upstream feedlot, or homeowners who cannot sell their property because they are downwind from the emission stacks of an incinerator.
- **Environmental Justice** – Poor, often minority people bear a disproportionate burden of most industrial environmental problems. Such lack of fairness based on racial, ethnic or socio-economic factors leads directly to justifiable outrage. With respect to the development, implementation and enforcement of environmental laws, regulations and policies, all people – regardless of race, national origin or income – receive fair treatment and are involved meaningfully when environmental justice is served.
- **Acts of God** – People are forgiving of 'acts of God,' yet extremely unforgiving of human-produced problems. Do not try to minimize problems you have by likening them to acts of God!
- **News Reports** – When a risk becomes news, outrage is likely to heighten. News media often report and amplify a problem that is rife with conflict and sensational anecdotes, while ignoring more deadly problems, which happen slowly or lead to more chronic consequences. Popular culture through films and television programs uses industrial catastrophes for dramatic fictional portrayals. These entertainment products get linked in the public mind to real possibilities.
- **Trust** – When facing a risk, people are more likely to stay calm if they trust the organization responsible for placing the risk on them. But, trust cannot be built in the face of an impending disaster. Organizational communications have to be proactive to be trustworthy. Better to always be sincere, respectful and courteous, than to try to build credibility in a crisis. Honesty is always the best policy.

A communicator's job may not always be to reduce outrage. Occasionally, the communicator's role may be to increase outrage. This may be a selected course of action when a risk exists but no one seems bothered about it. For instance, many times employees who work in high-risk occupations have regulations that mandate they wear specific equipment to protect them. If the equipment is bulky or the protocol required for full compliance is time-consuming, these workers may place themselves in harm's way by skirting safety precautions. Such a situation could be exposed by a cognizant communicator, and outrage encouraged.

If outrage exists where there is no demonstrable danger, only a false perception of it, then a communicator needs to address the outrage factors. Educational interventions may allow the audience to understand the true nature of the perceived risks. Conversely, if the hazard is real and outrage is low, then the communicator needs to educate about the full risk to prevent people from inadvertently harming themselves. One of the main sources of information used by the public to understand hazards is the mass media. A prime source of information generating outrage is word-of-mouth, though this can be mediated via telephone or email.

18.5 Mass Media Reports

Most people gain their information about what is dangerous from the news, though Internet-driven convergence has birthed the phenomenon of news-via-forwarded-emails. Science is rarely definitive, yet news media often express concern over the uncertainties in scientific data about risks. 'Experts' who disagree on interpretations of research are more newsworthy than their data, and fuel doubts and confusion in the non-scientific public. News accounts tend to give a simplified and erroneous view of the complexities of any risky situation. There are always uncertainties, but news media have not figured out a way to accurately report these yet.

18.6 Acknowledge Uncertainty to Communicate Risk Effectively

Communicating well about risk requires planning. A communicator needs to understand the science behind the hazard as well as the level and qualities of outrage being exhibited. The following questions help when trying to understand the hazard aspect of risk.

- Is there enough information to make a valid decision?
- What data are missing?
- What additional data are needed before a sound decision can be made?
- Are assumptions used for decisions made explicit in the interpretation?
- Are the scientific methods and statistics appropriate for the data used?
- Are all the data open for full scrutiny?
- Have all possible alternatives been considered?
- Have criteria for selection or rejection been clearly outlined and explained?

Having garnered information on the hazard, it is probable a lot of uncertainty still exists. Definitive answers are not always available. Yet the risk situation, even if high outrage and low hazard, must be dealt with and resolved if conflict is to be averted. Once an assessment is made of the situation, a plan should be drafted and used to manage the outrage. The goal is to help outraged people feel their concerns are being truly considered, while educating them about the real nature of the hazard. Of course, if the outrage is justifiably high, then the communication should be about how the hazard is being managed.

Risk is best discussed directly and in the open. An air of mistrust still lingers thickly around information about risk, perhaps an artifact of the Cold War and its many secrets. Audiences are more skeptical of messages about risk than any other subject.

Use pro-active public relations. Get interested citizens involved early. Deal with the situation's uncertainty and outrage head-on. Never ignore or dismiss outrage as unimportant.

Release news regularly. Do not suppress information, even if final data are not in-hand yet. Use community meetings. Develop an outreach program. Ask your audience about their concerns. Speak plainly, using minimal 'techno-babble.' Explain technical concepts. Be careful not to talk around the risk or to oversimplify it. Be cautious about the comparisons you use.

Develop a positive and orderly atmosphere where everyone can talk openly about risks, both hazards and outrage. Involve all affected stakeholders in decision-making. Listen to the concerns of all parties. Ensure even tangentially or potentially affected parties are identified and invited to take part. Even small groups can become a source of tremendous outrage when overlooked. Strive to create an inclusive dialogue. Do not take an authoritarian stance. Give all parties equitable standing in the decision-making process. If the corporation or government entity has veto power, so should the local citizenry.

State the limits of the data and admit uncertainty. Discuss data quality and confidence. Emphasize what has been done about a problem already and what is currently being done. Never cover-up information. Never lie. Be honest. Trust and credibility are absolutely essential.

Risks that frighten are not always the ones that do damage. While news media reporting in the past may have heightened concern for certain problems and issues, there remain many hazards that the public is mostly ignorant about. A risk communicator's job can be seen as being smart enough to discover situations when either hazard or outrage is dominating the risk equation, and then using their skills to inform those affected about the ignored portion of risk from the other side.

18.7 Final Thoughts

It has been said that we are safer and healthier than during any other period in history. If that is so, why are we more concerned about risk?

The last two centuries have seen the spread of industry and high technology around the planet. More technology, in general, engenders more risk. That is not to say that magnificent benefits – better health, longer life spans, higher quality of life, more entertainment opportunities, faster communication and transportation – have not come to modern humanity. Our greater reliance on new technologies has two important effects when considering the overall societal perception of risk. First, older technologies tend to become taken for granted. Machinery on family farms is some of the most hazardous technology around, but you hear few cries to save farmers from being mauled by their own plows and combines. Newer, unfamiliar technology breeds outrage, as people try to come to grips with its meaning in their lives. Think back over the last decade and the rapid development of wireless communications technologies. Have you heard urban legends about brain cancer and mobile phones?

The second factor our broadening technology base has had on our collective perception of risk is a statistical one. As populations have increased and the amount of technology available has risen as well, more accidents – in absolute terms – have been occurring. Even if each type of technology is dramatically safer, having hundreds or thousands more machines around still increases the number of problems that will occur. A one in a million eventuality can seem almost commonplace in a world of seven billion. Couple this numerical twist with a widespread belief that technology can be totally under our control and outrage germinates and spreads.

Experience with some sensationalized catastrophic mishaps of technology such as Three Mile Island, Chernobyl, and Bhopal have heightened our collective social concern to the point that anytime a bad thing happens we are ready to believe the worst possible report. We have become over-sensitized to industrial accidents.

A related social phenomenon is the confusion promoted by dueling experts in litigation. An incredible increase in litigation has caught more and more of the public's attention. They see hired guns shooting their mouths off, contradicting each other, all the while apparently using the same data. Credibility of science is eroded.

As we have become wealthier and healthier, it is seductive to feel we have more to lose. This fear can transform into paranoia, an obsessive and acute awareness of being surrounded by risks beyond one's control. To expect zero risk is to guarantee burning outrage. Pushing an image of a risk-free life is an open invitation to failure.

Mass mediated information from well-funded special interest groups plays a significant role in perceptions of ever-increasing danger. The message creation and delivery systems of interest groups have become more sophisticated. They can continually bring issues, both new and old, to our notice. And, the number of available channels available to spread information has increased exponentially. In short, we increasingly rely on others to tell us the truth. Indeed, it can be argued that we are so well-informed now from this broadening information base, that we become psychically paralyzed by the knowledge we possess. Wiebe (1973) gave this state of being the name 'well-informed futility.'

18.8   Case Study: Risk Analysis. Apples and Alar

A classic case of public relations and misinformation gone haywire is that of apples and the apple-ripening agent Alar. As early as 1973, Alar was shown to be capable of producing cancer in laboratory animals and eventually the U.S. Environmental Protection Agency gave the chemical a toxicity rating of '50 cancers per million adults from chronic long-term ingestion.' In 1989, Consumers Union tested Alar and ranked it with a lower toxicity of only five cancers per million. A Natural Resources Defense Council (NRDC) report stated the toxicity level was likely higher for children since their body mass was less than adults, but even then admitted an apple or glass of apple juice was healthier than candy that children might eat otherwise. Nevertheless, NRDC pushed forward with a campaign to ban Alar use in the apple growing industry, using high profile celebrities like actress Meryl Streep. News media scooped it up. A key point in heightened outrage was CBS's television's 'A is for Apple' expose on the highly rated and respected *60 Minutes* program, broadcast on Feb 26, 1989.

The applicable science got lost in the media translation, in favor of drama and anecdotes. Ultimately, the apple industry was forced to comply with questionable environmental regulations, which in hindsight may have been excessive in this case. In an ironic turn of coverage, news media later focused on the loss of markets for the apple growers.

For the next 6 years, the American Council on Science and Health led a campaign to show Alar as not only relatively safe, but also a victim of a scare campaign. Conservative think-tanks used Alar as an environmental 'Chicken Little' fable, to mark environmental activism as being harmful to American businesses. There then grew a public backlash, stymieing subsequent efforts by well-meaning consumer and environmental organizations, which were painted with the same alarmist brush for years to come.

Credit: Risk Analysis: Apples and Alav. Clipart 'Apple' Word 2009 Office.

References and Further Reading

Adubato S (2008) What were they thinking?: Crisis communication – the good, the bad, and the totally clueless. Rutgers University Press, Piscataway, NJ

Fearn-Banks K (2007) Crisis communications: A casebook approach, 3rd edn. Erlbaum, Mahwah, NJ

Gutteling JM, Wiegman O (1996) Exploring risk communication. Kluwer Academic, Dordrecht, The Netherlands

Hance BJ, Chess C, Sandman PM (1990) Industry risk communication manual: Improving dialogue with communities. Lewis Publishers/CRC, Boca Raton, FL

Hilldorfer J, Dugoni R (2004) The cyanide canary. Free Press, New York

Kamrin MA, Katz DJ, Walter ML (1995) Reporting on risk: A journalists handbook on environmental risk assessment. Michigan Sea Grant, Ann Arbour

Leiss W, Powell DA (2005) Mad cows and mother's milk: The perils of poor risk communication, 2nd edn. McGill-Queen's University Press, Montreal

Leviton LC, Needleman CE, Shapiro MA (1997) Confronting public health risks: A decision maker's guide. Sage, Thousand Oaks, CA

Lundgren R, McMakin A (2009) Risk communication: A handbook for communicating environmental, safety, and health risks. Wiley-IEEE, Hoboken, NJ

National Research Council (1989) Improving risk communication. National Academy Press, Washington, DC

Sadar AJ, Shull MD (eds) (1999) Environmental risk communication: Principles and practices for industry. Lewis, Boca Raton, FL

Sandman PM (1986) Getting to maybe: some communications aspects of siting hazardous waste facilities. Seton Hall Legislative J 9(2):437–465

Sandman PM (1991) Risk=hazard + outrage: A formula for effective risk communciation: Professional Development Courses and Products. American Industrial Hygiene Association Distance Learning, Fairfax, VA

Sellnow TL, Ulmer RR, Seeger MW, Littlefield RS (2008) Effective risk communication: A message-centered approach. Springer, New York, NY

Slovic P (1987) Perception of risk. Science 236:280–285

Society for Risk Analysis (2009) Risk analysis glossary. http://www.sra.org/resources_glossary.php Cited 27 July 2009

Ulmer RR, Sellnow TL, Seeger MW (2006) Effective crisis communication: Moving from crisis to opportunity. Sage, Thousand Oaks, CA

West B, Greenberg M (2003) Reporter's environmental handbook, 3rd edn. Rutgers University Press, New Brunswick, NJ

Wiebe GD (1973) Mass media and man's relationship to his environment. J Quart 50(426–432):446

Wogalter M, Dejoy D, Laughery KR (eds) (1999) Warnings and risk communication. Taylor & Francis, London

Chapter 19
Learning from Marketing and Public Relations

19.1 Introduction

Marketing and public relations are inventions of the business world for communicating messages to influence opinions of specific audiences. Popular opinions about these fields are often less than flattering, especially by those with environmental leanings. Indeed, a natural resources management student responded to a lecture on marketing communication with the comment that conservation was 'much too important to use marketing' in its pursuit. Furthermore, the concept of public relations is often thought of as putting 'spin' on bad situations to make them acceptable to the public.

In contrast, we believe both marketing and public relations represent powerful communication approaches which can be fruitfully, ethically and wisely used to deliver messages to audiences with respect to environmental issues. In this chapter, we appraise marketing and public relations and highlight how they can be used in environmental communication. Both are planned communication processes, and are quite similar to the communication planning model in Chapter 5. Social marketing, the marketing of ideas meant to promote change, especially can invigorate environmental communication campaigns.

19.2 Marketing and Social Marketing

The field of marketing is based on the concept of the market – a place where goods and services are exchanged. In most markets, participants exchange goods, services, money, information and even time. The concept of exchange, then, is critical to marketing. One can think of marketing as the managerial process by which people, or organizations, share information about what they would like to exchange. Most marketing texts distinguish marketing as a managerial process and a planned activity to understand an audience's needs and wants, and to find a means to fill these desires. A critical aspect of marketing is the concept of discovering unmet needs; marketing is not creating wants for products already in hand. Creating demand for items in hand is selling, not marketing.

A common example of marketing applied to a natural resources communication problem is solicitation of membership to a conservation organization. In order to raise money for the organization, they may offer enticements in exchange for your donation and affiliation. For example, a conservation organization may offer a tote bag, a magazine or a coffee mug for a certain size donation. To encourage larger contributions, enticements increase in value for larger donation sizes. Marketing is more than just having something to exchange. It is a planning approach based around four concepts: product, price, placement, and promotion.

Product is an item to exchange. In a marketing approach, a communicator tries to define for their audience a product that can be consumed. In the example above, organizational membership was defined by small premiums that were exchanged for membership fees. Members' dues will rarely generate enough funds for a conservation organization to operate. If the organization can infuse positive feelings of helping the environment into their products, then membership is far more powerful. Then, however, the organization must deliver: the member must be made to feel like they truly have helped.

Other products can be used in marketing approaches. In natural resource management, the resources themselves can be products. So, too, can privileges, such as the ability to hunt on a reserve or use of a campsite. A marketer always tries to think of value exchanges by defining consumable products.

Price is the amount an average person in an audience is willing to exchange for a product. Price is usually thought of as the amount of money that an audience exchanges for the product. But, price can also be non-monetary. For example, price can be time, bartered goods, opportunity costs and peripheral costs used in an exchange. When using a marketing approach, the communicator evaluates the entire cost to their audience, not just the amount the 'seller' will receive. For example, a community comparing options of curbside recycling versus a centralized recycling center would include costs of participant travel to a recycling center in the comparison, including gasoline and travel time.

Placement refers to distribution of a product. A marketer must answer the question 'How can the audience obtain the product?' Many environmental groups solicit memberships by mailing promotional materials to perspective members' homes, thus making for convenient placement. A new member need only write a check or give a credit card number, and mail a reply back to the organization. The same exchange may take place as an on-line exchange too. As another example, hunting and fishing licenses are often sold by state departments of natural resources through sporting goods stores, which are places that hunters and anglers are likely to visit anyway.

Promotion refers to the manner in which an audience is made aware of a product, as well as its price and placement. Advertising and personal selling are two ways, among others, that promotion is achieved. Promotion may also refer to a message sent about a product: is it healthy, environmentally-friendly, or money-saving? An audience is provided information to make a decision to obtain a product. As an example, some states have tax check-off programs to fund natural areas protection; these are often promoted through posters, special mailings and special banners on agency web sites.

Marketing, then, is the application of a planned process to meet wants and needs of an audience. Social marketing is the application of these principles to a goal of achieving some sort of social change (Zaltman et al 1972). The product becomes the particular social change itself, while the price is what individuals give up to achieve the change. A fifth 'P', as defined by Zaltman et al (1972), place is 'outlets which permit translation of motivations into actions.' This gets at the idea that while people consider themselves environmentalists, they do not have the opportunity to act on their expressed ideals. A marketing approach to environmental behavior must address this need for actionable options. Promotion, in a social marketing perspective, is the raising of awareness of the desired change.

Marketing and social marketing can be powerful tools for communicators. When resource managers adopt a product orientation, they may find new and more effective means of achieving objectives. Certainly traditional natural resources, such as forest products, have been approached through marketing. Many other resource problems can be addressed through product orientation: interpretive services, recreation, fundraising and volunteer recruitment. Objectives specifying behavior change, such as recycling, green consumerism, saving energy, eliminating poaching and so forth can also be addressed using a marketing approach. This is an important component of the environmental communicator's toolbox. At first, using social marketing techniques will feel highly creative. With time and success, such skills become routine as well as valued.

19.3 Public Relations

The field of public relations is often maligned. The phrase can connote communications designed to 'spin' unacceptable events so an enraged public will swallow them and quell. This is probably caused by the use of public relations strategies after some negative event has occurred to try to mitigate its impacts. Many managers only think of public relations reactively, after a public is fuming over some event or action. Public relations, when used proactively, can be a powerful, two-way communications tool for an organization to establish trust and mutual benefits with its publics.

The Public Relations Society of America uses the following vision statement: 'Public relations helps an organization and its publics adapt mutually to each other' (PRSA 2009). This indicates how public relations ideally creates a dialogue between an organization and their community. Indeed, many 'PR disasters' would not occur if sufficient public relations work was done as an ongoing part of business.

Several specific types of public relations are practiced by organizations. Each of these uses the same principles, but has different goals and specific activities. All of the types of public relations described below are common in environmental fields.

Media relations consists of establishing and maintaining working contacts with news media professionals, including those in local and national newspapers, television, radio, bloggers and other mass media outlets. In media relations, it is important to understand that credibility and trust must be established at two levels: first with journalists, and second with the final audience. Relationships can be fostered with reporters and editors through careful preparation of news releases and offering frank, honest interviews, demonstrations and tours. Exclusive stories are valuable, so offering an interview to a single outlet can increase its probability of being used as well as it prominence, and can build goodwill with the chosen journalists. It is important when working with news media to understand the needs of reporters and editors (Chapter 16).

Government relations are activities meant to influence policy making (law making and rule enforcement) by government organizations. Personal actions such as lobbying are often used, but use of media and mass mobilization campaigns are common supporting or primary activities of government relations. This area requires not only an understanding of the principles of public relations, but also knowledge of how government works and regulations concerning lobbying.

Community relations include activities to include local stakeholders in planning and decision-making. Even when companies, governments or nonprofit groups have no legal obligation to work with local residents, good community relations can lead to creative problem solving and reduce the likelihood of resistance to actions. Besides being a good neighbor, organizations can practice open, honest dialogue within their community, both listening to concerns and providing meaningful, appropriate information. The principles of risk communication (Chapter 18) and conflict management (Chapter 17) are important in this area of public relations.

Crisis communications are those public relations activities used when disasters occur. Crises can involve corporations, such as the *Exxon Valdez* oil spill; governments, such as explosions of space shuttles *Challenger* and *Columbia*, and even nonprofit organizations, such as too-common embezzlement scandals. Organizations should prepare crisis communication plans in advance, by determining which kinds of crises might occur, and preparing communication plans in advance, including designating who should be on the communication team. Even an unanticipated crisis should be considered with a generic communication plan, a basic tenet of which is 'Who speaks for the organization when it is in crisis?' As crises are inevitable, organizational response can save lives, ecosystems, reputations and money.

Internal communications are those for which the target audience is employees, members or volunteers of the organization. These messages are designed to keep internal audiences informed and motivated. Media include printed items such as newsletters, electronic media such as intranet sites, employee-only presentations and 'town hall meetings,' and one-on-one discussions. Like all public relations, open, honest, two-way communication is critical. It is important to remember the boundary between the organization and external audiences is permeable: employees hear external messages, and external audiences eventually hear internal messages.

Public relations properly conducted, according to PRSA (2009), benefits both the organization and society. It allows the organization to be responsive to its publics, thus making it more effective. At the same time, public relations informs the organization about its social responsibility by giving voice to community constituencies. The key to effective, meaningful public relations is the same as other communication programs – planned communication to specific audiences with many opportunities for feedback.

19.4 Propaganda

Marketing and public relations are often accused of practicing propaganda. Propaganda can be thought of as lying by telling the truth – presenting some facts while obfuscating or ignoring others to present a distorted view of reality. Certainly, marketing and public relations messages, as well as many environmental messages, are intended to present a point of view. Pratkanis and Aronson (2001) stated, 'Every day we are bombarded with one persuasive communication after another. These appeals persuade not through the give-and-take of argument and debate, but through the manipulation of symbols and of our most basic human emotions. For better or worse, ours is an age of propaganda.' This observation prompts an important caution for environmental communicators, who often feel a moral commitment to the messages they craft. Propaganda used in pursuit of a social good is still propaganda.

Still, there is a distinction between persuasive messages and propaganda. This distinction fails when dissenting voices are not allowed to speak. Environmental communicators must acknowledge other points of view. Environmental benefits will not be sustained if some publics are left out of the communication process.

19.5 Greenwashing

Modern lifestyle can be attributed to the monetary and technological wealth that grew out of the Industrial Revolution. One of the most common admonitions against business is that in the manufacturing so many products, a legacy of toxic pollution and hazardous by-products has also been created. Industrial pollution has led to myriad environmental issues that confront us today. Without laying blame which is no easy task, let us focus on one of the communication perspectives that confront business – the use of 'green-marketing' to sell products.

John Grant (2008) emphasizes we live in an age where 'green' is perhaps the leading marketing strategy being employed by the business community. But, does green marketing strategy actually deliver on sound green goals and objectives, as opposed to mere green theming? It takes less effort to make a product look green, than to manufacture a truly green product. As communicators we need to be careful of painting all corporate marketing as *greenwashing*, which is akin to propaganda where something is cast to look green when it really is not. As a communicator, one develops skills to sift through the environmental messaging debris to understand how corporations are improving, or abusing, the environment. As an example, the environmental activist group Greenpeace launched a web site, http://www.stopgreenwash.org/, as their way to combat this particular problem. There are companies that are really trying to be green – yes, there are many out there. We should emphasize that many different groups engage in greenwashing besides businesses. Anyone with an agenda to look green, such as government agencies, politicians, NGO groups with special interest agenda's, etc., may utilize greenwashing as a way to improve their image within public perceptions of what being green really means. Critical thinking and the guidelines below can help us identify those that are really 'walking the talk' from those that are merely creating smokescreens to obscure their real intentions.

Sourcewatch, a wiki-based project of the Center for Media and Democracy, developed some tips for detecting greenwash. Big budget green-washing corporate campaigns are designed to obscure so as to defuse skepticism of journalists, politicians and activists. Some rough rules of thumb for testing whether the claims made by a company, government or even a non-governmental organizations are genuine (Sourcewatch 2008):

- **Follow the Money Trail**: Many companies are donors to political parties, think tanks and other groups in the community. Not all for-profit companies disclose in their annual reports who they donate to, even though it is shareholders' money. Ask about all their donations, not just those they boast about in glossy documents such as corporate social responsibility reports.
- **Follow the Membership Trail**: Some companies boast about virtues of their environmental policy and performance, but hide anti-environmental activism behind the banner of an industry association to which they belong. Find out what industry association companies are members of and check the association policies. Watch for paid placement of stories via journalists and funding of university-based research. Corporations, through industry groups or public relations firms, also like to sponsor so-called think tanks. Corporate-sponsored front groups are used to appear independent, so that industry's words come out of the mouths of others (called *third party technique*).
- **Follow the Regulatory Paper Trail**: Most companies, and their associated trade groups, make submissions to governmental units on a wide range of issues. These submissions are public information, and many are now posted on searchable databases. When corporations and trade groups send letters to politicians and government agencies, those become public information as well. In the United States, public information can be accessed by Freedom of Information Act searches. More than 80 democracies have similar legal access guarantees.
- **Look for Skeletons in the Company's Closet**: Some companies include information in annual reports about problems that have been in the news in the last year. But, there may have been other problems, occasionally reported in the news, which they do not want to tell shareholders about. Check for information on the company with watchdog groups and in searches of the news, and compare that with what they disclose.
- **Test for Access to Information**: Many companies make lofty claims about their commitment to transparency and providing information to stakeholders. Do not just take them at their word. In reports, they will probably refer to environmental impact statements, reviews, audits, monitoring data and other information.

If it relates to an issue you are interested in, ask to see the primary source. Remember that 'commercially confidential' is corporate-speak for 'no.'
- **Test for International Consistency**: Companies may operate to different standards in different countries. Check whether operating standards and procedures are consistent across national boundaries, or whether they opt for lower standards where they think they can get away with it.
- **Check How They Handle Critics**: Some companies go to extraordinary lengths to try and silence their critics. This can involve everything from legal threats to questionable corroboration with sympathetic police and military forces.
- **Test for Consistency Over Time**: An unscrupulous company will launch a green initiative with much fanfare, and then starve it of funds. Keep watching when the spotlight fades.

19.6 Philanthropy as Communication

The last tip for spotting greenwash indicates the overarching importance of funding. Without financial resources, few environmental campaigns succeed. That 'money talks,' as the adage goes, shows the communicative power that comes when dollars are devoted to a cause.

Fundraising is a never-ending function for most environmental non-governmental organizations, and provides fuel for their programs. Corporations certainly know this, and use it to inform their philanthropic decisions. 'Corporate philanthropy can be viewed as an extension of the corporate communication or public relations function,' noted Genest (2005). When a philanthropic gift is made as part of an environmental campaign, it adds a gloss of promotion. Funding may accelerate and amplify a campaign. Because a gift is also an exchange, philanthropy can be savvy social marketing for both giver and receiver.

Many non-governmental organizations go through serious handwringing when offered a corporate gift. On the recurrent question of 'clean versus dirty money,' the best advice we have heard comes from Klein (2007). At an appropriate staff or board meeting announce the offer of the gift and the potential corporate donor. Gauge the immediate reaction of the group. If there is a visceral response akin to nausea, rather than smiles and nods, the gift has not passed Klein's 'gag test.' She suggests it probably should be declined.

19.7 Summary

Marketing, especially social marketing, and public relations are powerful tools for environmental communicators. Both are planned approaches to communication to achieve specific objectives using certain techniques. Marketing and public relations approaches can be highly effective, so we encourage their creative application to a wide variety of natural resources settings.

19.8 Case Study: Marketing and Public Relations. Organic Food Is Harmful?

In 1998, Dennis Avery of the Hudson Institute (a free-market conservative think tank) made a startling claim that eating organic food was more dangerous than eating food produced using chemical pesticides. He attributed this claim to the use of animal manure, which he stated had high levels of *E. coli* bacteria – the virulent strain O157:H7 strain specifically. Avery claimed validation came from Paul Mead, an epidemiologist at the U.S. Centers for Disease Control. Mead countered the data that Avery attributed to him was bogus and he had nothing to do with them. This repudiation did not stop Avery though. Indeed, Avery's statement that organic food is eight times more likely to kill you because it is 'dirtier' keeps echoing in stories about organic foods. Counterclaims came from Fred Kirschenmann of the National Organic Standards Board and Robert Elder of the U.S. Department of Agriculture's Meat Animal Research Center, both specialists in O157:H7 contamination. They counter feedlots are much more likely to harbor the deadly bacteria than pastures. Despite frequent debunking, Avery's myth persists. The use of junk science and misinformation to push a false message is combated with a valid and proactive public relations counter-campaign. As this case shows, a false message will not just go away.

Credit: Marketing and Public Relations: Organic food is harmful? Clipart 'Food' Word 2009 Office.

19.9 Case Study: Marketing and Public Relations. U.S. Environmental Protection Agency and 9/11 pollution

When a crisis forces an agency into a reactive stance, public perceptions must be dealt with as real. The September 11, 2001, attacks on New York City put the U.S. Environmental Protection Agency (EPA) into just such circumstances. In the early aftermath, Christine Todd Whitman, then head of EPA, made the classic faux pas of dismissing valid concerns about toxicity within dust and debris from the fallen World Trade Center towers. In some of her first public statements, Whitman failed to acknowledge there was any asbestos or other toxins in the dust blown around the city after the towers collapsed. This lack of concern continued for at least a further 14 months, despite a growing amount of evidence to the contrary. Multiple rescue and salvage workers developed severe respiratory problems from working at the site. Instead of honesty, straight talk and acceptance that the dust concerns were real, Whitman kept dismissing concerns, hid behind legal counsel, and tried to imply she had not been hiding any facts about toxicity in the dust. She maintained an adversarial and defensive posture throughout, which served to anger victims and their families. The opposite approach would have been a willingness to share facts, an empathic and approachable demeanor, and an aim to assist.

Credit: Marketing and Public Relations: U.S. Environmental Protection Agency and 9/11 pollution. Clipart 'Agencies' Word 2009 Office.

References and Further Reading

PRSA (Public Relations Society of America) (2009) http://www.prsa.org/aboutUs/officialStatement.html. retrieved July 2, 2009

Andreasen AR (1995) Marketing social change: changing behavior to promote health, social development, and the environment. Jossey-Bass, San Francisco, CA

Bland M, Theaker A, Wragg D (2005) Effective media relations: how to get results, 3rd edn. Kogan Page, London

Boone LE, Kurtz DC (2008) Contemporary marketing, 14th edn. South-Western College, Chula Vista, CA

Broom G (2008) Cutlip and Center's effective public relations, 10th edn. Prentice Hall, Upper Saddle River, NJ

Butler T (ed) (1998) Wildlands philanthropy. Wild Earth 8(2):12–56

Butler T, Vizcaino A (2008) Wildlands philanthropy, the great American tradition. Earth Aware Editions, San Rafael, CA

Caywood CL (ed) (1997) The handbook of strategic public relations and integrated communications. Mcgraw Hill, New York, NY

Churchill GA, Iacobucci D (2009) Marketing research: Methodological foundations, 10th edn. South-Western College, Chula Vista, CA

Dilenschneider RL (2008) Dartnell's public relations handbook, 4th edn. Dartnell Corporation, Chicago, IL

Fazio JR, Gilbert DL (2000) Public relations and communications for natural resource managers, 3rd edn. Kendall Hunt, Dubuque, IA

Genest CM (2005) Cultures, organization and philanthropy. CCIJ 10(4):315–327

Goldberg ME, Fishbein M, Middlestadt SE (1997) Social marketing: Theoretical and practical perspectives. Erlbaum, Mahwah, NJ

Grant J (2008) The green marketing manifesto. Wiley, New York, NY

Harrison EB (2008) Corporate greening 2.0. PublishingWorks, Exeter, NH

Hawken P (2008) Blessed unrest: How the largest social movement in history is restoring grace, justice, and beauty to the world. Penguin, New York, NY

Hendrix JA, Hayes DC (2009) Public relations cases, 8th edn. Wadsworth, Belmont, CA

Kendall R (1996) Public relations campaign strategies: Planning for implementation. Harper Collins, New York, NY

Klein K (2007) Fundraising for social change, 4th edn. Jossey-Bass, San Francisco, CA

Kotler P, Lee NR (2002) Social marketing: Influencing behaviors for good. Sage, London

Kotler P, Roberto N, Lee NR (2002) Social marketing: Improving the quality of life. Sage, London

Malhotra NK (2009) Marketing research: An applied orientation, 6th edn. Prentice-Hall, Upper Saddle River, NJ

Moser SC, Dilling L (2007) Creating a climate for change: Communicating climate change and facilitating social change. Cambridge University Press, Cambridge

Pratkanis AR, Aronson E (2001) Age of propaganda: The everyday use and abuse of persuasion. Holt, New York, NY

Sourcewatch (2008) Greenwashing. Sourcewatch encyclopedia. http://www.sourcewatch.org/index.php?title=Greenwashing. Cited 28 July 2009
Tuten TL (2008) Advertising 2.0: Social media marketing in a web 2.0 world. Praeger, Westport, CT
West B, Greenberg M (2003) Reporter's environmental handbook, 3rd edn. Rutgers University Press, New Brunswick, NJ
Zaltman G, Kotler P, Kaufman I (eds) (1972) Creating social change. Holt, Rinehart & Winston, Austin, TX

Chapter 20
Walking the Talk of Green Business and Sustainability

20.1 Introduction

Corporations have often been characterized as perhaps the biggest contributor to environmental problems. But, signs of a transformation can be observed (Edwards 2005). Business sector response to this transformation is linked by communicators to empowerment for communities. Communities and businesses work together to create their own forms of sustainable development. Stakeholders in regenerative communities display appreciation for the different ways of thinking that occur in business and sustainability. Regenerative communities manage to encompass all stakeholders.

In Chapter 1, quality of life, the Triple Bottom Line and the Quad Stack were shown as basic principles assessing what needs to be done by the business community to attain sustainability. Ideally, businesses operate with ecological literacy as a core function of their business strategy. In reality, this is easier said than done as the ethical framework for making this happen is poorly applied.

Research shows that business organizations, like individuals residing in a common region, build a shared concept of place through their actions, values, behaviors and strategies (Thomas and Cross 2007; Thomas et al. 2009). The social construction of place from an organizational perspective involves balancing two requisites:

- Resource dependency between the organization and the community, as each seeks mutual advantage
- The role of the organization as a responsible agent that advocates for and protects its place

Such a holistic view helps to frame the organization as a critical agent accountable for and to its place. Thomas (2004) identified four qualitatively different types of organizations, each of which values place in unique ways and therefore varies in how it contribute to or detracts from social construction of place. These four are:

- Transformational
- Contributive
- Contingent
- Exploitive

They are differentiated along two dimensions, their agent perspective (how they view themselves as a member of a place) and their organizational mission (whether they are committed to being an interdependent member of the place or just to exploit it).

Smaller businesses are more likely to be demonstrably green and they do so voluntarily and with less publicity, compared to transnational corporations who are more likely to promote actively their sustainability practices, possibly even through greenwashing. Forces pushing organizations to be sustainable, as well as how they might become agents for community transformation is the subject of this final chapter. Are they 'walking the talk'? Or, just 'washing themselves green'?

20.2 Corporate Social Responsibility

In the ideal, socially responsible corporations volunteer to behave ethically in all their business dealings, with 'societal good' a major goal of success. In a bottom-line, profit-based and highly competitive setting, the usual reality is that most businesses adopt societal good as a goal only under heavy external pressure. So, they act in environmentally responsible ways when it is good for their immediate competitive advantage and when they are being watched intently.

Deborah Doane (2005) critiqued sustainable corporate business through four myths:

The market can deliver both short-term financial returns and long-term social benefits: Interests of profit-seeking corporations and the broader society are often at odds. Socially responsible investments by corporations are unlikely to pay off in the short time horizon demanded by stock markets. A solution would be to help shareholders see the benefits of long-term strategizing for sustainability.

The ethical consumer will drive change: Survey after survey shows consumers are more concerned about price, taste or sell-by date than corporate social responsibility. The solution here is to explain the role of over-consumption in environmental degradation, which has been an emphasis of environmental education for many years. Such educational messages have to push through awareness to empower consumers to change their habits. As noted in Chapter 12, we are in a transition worldview phase, collectively creating a new vision for change.

There will be a competitive 'race to the top' over ethics amongst businesses: Doane wrote, 'While [corporate social responsibility] efforts often offer good [public relations], which companies of course like, in some cases businesses may be able to capitalize on well-intentioned efforts, say by signing [onto major initiatives for sustainability] without necessarily having to actually change their behavior.' Such corporate behavior is a double-edged sword because businesses are hesitant to change when consumers and/or shareholders are not fully supportive of 'green' initiatives, yet support of these initiatives makes businesses look more compassionate.

In a global economy, countries will compete to have the best ethical practices: 'Although companies often claim that their presence in developing countries will improve health, environmental and labor conditions, companies often fail to uphold voluntary standards of behavior in developing countries, arguing instead that they operate within the law of the countries in which they are working. In fact, competitive pressure for foreign investment among developing countries has actually led to governments limiting their insistence on stringent compliance with human rights or environmental standards, in order to attract investment,' Doane countered. In a nutshell, investing in a Triple Bottom Line or Quad Stack philosophy can mean cutting bottom-line profits.

While these myths initially sound demoralizing for the prospects of real and lasting change, many environmental-minded businesses and environmental non-profit groups have drafted frameworks for sustainable operation of businesses. They begin to sketch the picture of how to thrive in a post-modern world where sustainability is a prerequisite for human success.

20.3 Frameworks for Sustainable Business Practices

Since the basic ideas of sustainability are broad – the interrelatedness of all life, including humans, and the reliance on all life on Earth's natural resources and ecological services – you will see common themes throughout the frameworks. The following frameworks were created by organizations promoting principles of sustainability within a business perspective. As the rise of the business world has generated a more urban human population, it should come as no surprise that some guidelines focus on the architecture of the built environment. We give only a few of the more popular approaches. There are numerous others, an indication of the traction this way of thinking has gained. The business sector's deciphering of the practice of sustainability will most assuredly affect all our lives at all levels.

20.3.1 Hannover Principles

In 1992 as part of the planning for a world's fair in Hannover, Germany, McDonough and Braungart, co-authors of *Cradle to Cradle: Remaking the Way We Make Things* (2002), conceived a set of business principles with sustainable design (designing buildings and objects with forethought about their environmental impact) as a basis for future manufacturing. The nine Hannover principles are (McDonough 1992):

- Insist on rights of humanity and nature to co-exist – while it may seem a revolutionary idea that nature should have rights, the notion has its modern roots in the Aldo Leopold's *Land Ethic* (1949).
- Recognize interdependence – be ecologically literate.
- Respect relationships between spirit and matter – not just an abstraction about humankind's ultimate nature and purpose, this also concerns our keystone role as dominating, tool-wielding organisms in the biotic-abiotic world of interdependent communities.

- Accept responsibility for the consequences of design – recognize our current consumer lifestyles cause negative systemic changes on the planet. A new form of sustainable consumerism can arise from a redefinition of social needs and wants.
- Create safe objects of long-term value – stop designing for planned obsolescence.
- Eliminate waste – mimic natural systems where there's complete recycling of everything.
- Rely on natural energy flows – mimic natural systems of energy capture and use, and stop using fossil fuels.
- Understand the limitations of design – some products' industrial manufacture is just not meant to be; they need a complete rethinking of how they can work.
- Seek constant improvement through the sharing of knowledge – let's communicate effectively.

20.3.2 Sanborn Principles

In 1994, the National Renewable Energy Laboratory, a U.S. Department of Energy facility, convened a wide-ranging group of experts to explore sustainability. Their goal was to develop a set of guidelines for those wishing to pursue sustainable development. To do this, they imagined an idyllic city wholly designed for sustainability, which they named Sanborn. From their visions and designs for Sanborn came these principles (Harwood 1997):

- **Ecologically responsive**: This principle recognizes ecological limits and patterns of natural ecology within specific areas, which dictate the least disruptive land use and building designs. New urban systems should concentrate where facilities are most accessible and 'walkable.'
- **Healthy, sensible buildings**: All human habitat designs should create a healthy and holistic living environment. Buildings should be made from organic materials, integrate art, allow sunlight in, contain green plants, be energy efficiency, have low noise levels, and conserve water. Features of buildings and their surroundings should include:
  - No waste that cannot be assimilated – e.g., use of living systems for waste water and sewage treatment.
  - Thermal responsiveness – use of effective thermal insulation, natural ventilation, and passive and active solar design.
  - Reflective or actively productive roofing and parking surfaces – design large surfaces to be thermal exchanges or energy collecting systems.
  - 'Junglified' or planted with native vegetation, both exterior and interior.
  - Access by foot to primary services – reduce need for unnecessary vehicular use.
  - Natural corridors near residences for wildlife.
  - Individual and/or community gardens, including local agriculture for local consumption.

- **Socially just**: Habitats have equal access across economic classes.
- **Culturally creative**: Habitats allow ethnic groups to maintain individual cultural identities and neighborhoods, while integrating into the larger community with many cultural events.
- **Beautiful**: Beauty in a habitat-based environment is necessary for the soul development of human beings.
- **Physically and economically accessible**: All sites within the habitat shall be accessible and rich in resources to those living within walkable (or 'wheelchair-able') distance. Accessible characteristics shall include:
  - Motor vehicle traffic calming
  - Clean, accessible, economical mass transit
  - Bicycle paths
  - Small neighborhood service businesses; i.e. bakeries, tailors, groceries, fish and meat markets, delis, coffee bars etc.
  - Places to go where chances of accidental meetings are high; i.e. neighborhood parks, playgrounds, cafes, sports centers, community centers etc.
- **Evolutionary**: Continual evaluation of premises and values. Demographically responsive and flexible over time to support future user needs. This societal heterogeneity should be:
  - 'Villagified' – should be discrete, yet pluralistic smaller communities that inter-relate to surrounding communities
  - Multigenerational
  - Non-exclusionary

20.3.3 *Principles of Ecological Design*

Van der Ryn and Cowan (2007) helped pioneer five principles of ecological design that reflect much of the spirit within the Sanborn Principles.

- **Solutions grow from place**: Because local conditions and surroundings matter, we need to think about solutions as they apply to individual locales. Design based on local climate and vegetation and build with local materials.
- **Ecological accounting informs design**: Create and use new metrics to measure the full environmental impacts of existing or proposed designs, so as to determine the most ecologically sound designs. This task demands sound ecological literacy.
- **Design with nature**: Utilize ecosystem services and work well within their limitations.
- **Everyone is a designer**: Honor the special knowledge each person brings to a project.
- **Make nature visible**: Reveal ecological processes within human-built structures. Natural environments are more pleasing, reduce stress and facilitate more community.

20.3.4 Leadership in Energy and Environmental Design

Leadership in Energy and Environmental Design (LEED) is an internationally recognized certification system to measure how well a building or community performs across several metrics that affect environmental quality: energy savings; water efficiency; CO_2 emissions reduction; improved indoor environmental quality; stewardship of resources and sensitivity to their impacts; and innovation and design processes (U.S. Green Building Council 2009).

LEED was developed beginning in 1994 by a broad group of non-profit representatives, government agents, architects, engineers, developers, builders, product manufacturers and other industry leaders (Scheuer and Keoleian 2002). A pilot certification program was released in 1998. Operated by the U.S. Green Building Council, LEED has become the defining certification system for ecologically sound building practices. Both new and retrofitted buildings can be LEED certified. LEED's primary goals are:

- Define 'green building' by establishing a common standard of measurement.
- Promote integrated, whole-building design practices.
- Recognize environmental leadership in the building industry.
- Stimulate green competition.
- Raise consumer awareness of green building benefits.
- Transform the building market.

The third version of the LEED certification (U.S. Green Building Council 2009) has 100 possible base points plus an additional 6 points for innovation and 4 points for underserved regions. Scoring is subject to third-party verification. Once scored, buildings can qualify on four levels of LEED certification:

- **Certified** (40–49 points)
- **Silver** (50–59 points)
- **Gold** (60–79 points)
- **Platinum** (80 points or more)

LEED values thoughtful design of human habitats to promote more environmentally friendly habits.

20.3.5 A Sense of Place for Businesses

The remaining principles move away from design of the built environment, to relate more directly to business practices and strategies. When applied, these principles shape corporate social responsibility and involve core ideas inherent in the Quad Stack.

Amy Townsend (2006) succinctly stated five basic strategies in creating a model for an environmentally responsible company:

- **Greening your mission**: Making a public commitment is the first step in changing how a business functions.

20.3 Frameworks for Sustainable Business Practices

- **Greening your employees**: Communicating with and educating employees in both sustainable practices and how those practices affect their specific jobs is a crucial step in transformational leadership.
- **Greening your operations**: Once there is buy-in from all corners of the business, the operations can be streamlined and 'greened' for real efficiency using full accounting practices.
- **Greening your facilities/sites**: Auditing the building for energy savings and more effective energy co-generation. Generally, life-cycle analysis of the businesses materials will unveil better waste stream management and full accounting.
- **Greening your products/service with green practices to change industry**: Really walking the talk.

Businesses already moving in these directions show cultural changes, both in their strategies for corporate social responsibility and the practices they use as they approach sustainability. Examples are:

- **Patagonia**: In its mission statement, the outdoor clothing and gear manufacturer commits to, 'Build the best product, cause no unnecessary harm, use business to inspire and implement solutions to the environmental crisis.' In a related corporate social responsibility commitment, they are 'Dedicated to promoting fair labor and environmental protection where Patagonia products are made.' Besides facilities and process improvements, Patagonia now lists 'Product Footprint' data for a large portion of their products, to help consumers make more informed purchasing decisions based on information about the ecological and social costs of products.
- **Interface Inc**: Their vision statement is quite explicit: 'To be the first company that, by its deeds, shows the entire industrial world what sustainability is in all its dimensions: People, process, product, place and profits.' Their lengthy mission statement is in the same vein: 'Interface will become the first name in commercial and institutional interiors worldwide through its commitment to people, process, product, place and profits. We will strive to create an organization wherein all people are accorded unconditional respect and dignity; one that allows each person to continuously learn and develop. We will focus on product (which includes service) through constant emphasis on process quality and engineering, which we will combine with careful attention to our customers' needs so as always to deliver superior value to our customers, thereby maximizing all stakeholders' satisfaction. We will honor the places where we do business by endeavoring to become the first name in industrial ecology, a corporation that cherishes nature and restores the environment. Interface will lead by example and validate by results, including profits, leaving the world a better place than when we began, and we will be restorative through the power of our influence in the world.' Interface further qualifies these ideals with a production focus branded as 'Achieving Mission Zero,' in which Interface promises to eliminate any negative impact by 2020.

20.3.6 Corporation 20/20

Corporation 20/20's central framing question is 'What would a corporation look like that was designed to seamlessly integrate both social and financial purpose?' This multi-stakeholder initiative seeks to answer the question and to communicate about new systemic corporate designs where social purpose is a primary concern. They offer six principles (Corporation 20/20 2008):

- The purpose of the corporation is to harness private interests and to serve the public interest. This is essentially the core of capitalism that was developed by the early economic philosopher Adam Smith, who espoused 'enlightened self-interest' for the social good and not selfishness as is often attributed to his 'invisible hand' metaphor.
- Corporations shall accrue fair returns for shareholders, but not at the expense of the legitimate interests of other stakeholders. Profits are OK, but not if they cause any distress or hardship to others.
- Corporations shall operate sustainably, meeting the needs of the present generation without compromising the ability of future generations to meet their needs. Natural resources shall not be exploited, because industrial ecology follows rules of natural ecological systems.
- Corporations shall distribute their wealth equitably among those who contribute to its creation. All investors, buyers, sellers and workers that produce goods should benefit equitably from all manufacturing. Accounting includes externalities such as impacts on ecological systems. All goods should be fully priced to include all externalities.
- Corporations shall be governed in a manner that is participatory, transparent, ethical and accountable.
- Corporations shall not infringe on the right of people to govern themselves, nor infringe on other universal human rights.

Too often the debate about the future of corporations is presented as an absolute choice between either heavy government regulation or unrestricted free markets. Corporation 20/20 posits a third path: redesign of the corporate system. Influential business thinkers like Paul Hawken (1994, 1999) and Amory Lovins of the Rocky Mountain Institute have advocated the same concept for years. Reaching beyond mainstream implementations of corporate social responsibility, Corporation 20/20 tries to 'chart a path that embeds social purpose in the organizational genetics of corporate structure while helping to build high-performing organizations.'

20.3.7 *Green to Gold*

Green to Gold: How Smart Companies Use Environmental Strategy to Innovate, Create Value, and Build Competitive Advantage, by (Esty and Winston 2006), is a 'hard-knocks' look at trying to attain sustainability in business. Like a reality TV show, it peers in on doing the right thing, but then shows what really happens and what has not worked within the modern business paradigm. One of the major lessons is that droves of consumers do not yet seem ready to make green choices enthusiastically. They outline a series of steps to become green within the limitations still imposed by lack of purchaser commitment, but their big message is that 'Truth Matters.' To business that means no greenwashing. Purchasers, they note, are slowly making a change. An interesting feature is the inclusion of a table that lists 'wave-riders' – the top twenty companies in the United States and internationally that are striving to become sustainable companies. These are the business leaders atop the front wave of the sustainability revolution.

20.4 Thinking Differently, Thinking Systemically

Business commitment to sustainable thinking is progressing, albeit slowly. Several experts promote ideas in how to successfully make the transition to sustainability. One of the main gurus for systemic thinking is Peter Senge (2006, 2008). He has three guiding ideas to support his work:

- There can be no viable path that does not incorporate intergenerational thinking
- Social-cultural institutions matter
- Any real change must be grounded in new ways of thinking

As Albert Einstein said, 'No problem can be solved with the same mindset that created it.'

Senge's prerequisites for systemic thinking are seeing smaller systems within larger systems; recognizing that many minds from different disciplines are needed to think of systems; and, connecting and adjusting instead of problem-solving in isolation. We need to convene a diversity of viewpoints, be open to all options without specific advocacy except for sustainability, and to nurture relationships above money. This set-up then links to six basic themes:

- The natural system encloses social and economic systems.
- An industrial system must operate at least in the context of the Triple Bottom Line.
- All potentially renewable resources have harvest limits.
- All non-renewable resources are finite.
- Thinking of waste is pathological.
- Socio-cultural community is the vessel for change.

A major question for many enlightened consumers is how to know when a company is really walking the talk. A company may not be resorting to greenwash tactics, but still be as brown as they come. What about companies which are serious about their commitment to sustainable principles?

20.5 Corporate Sustainability Reporting

In the 1980s, in light of environmental action and regulations fostered during the 1970s, many companies were finding their negative corporate images were a cause of lost profits. During this time, greenwashing first becomes highly noticeable, especially within the chemical industry. It is also during this period that environmental management systems (EMAS) were being adopted, led by European companies. By the 1990s, a few of these corporations also began furnishing EMAS statements open to public scrutiny. By the end of the 1990s, there was a large number of companies, European especially but a few U.S. ones as well, reporting their impacts (good and bad alike) through what has come to be called corporate sustainability reports.

The Global Reporting Initiative (GRI) provides the world's de facto standard in sustainability reporting guidelines. A sustainability report to GRI standards is public, usually annual, and provides data and analysis concerning economic, environmental and social impacts. Most often these are related in Triple Bottom Line or Quad Stack fashion. The GRI goal is to have all organizations report their sustainability (or lack thereof) as readily as they do their finances. More than 1,500 organizations in 60 countries have produced a sustainability report (Global Reporting Initiative 2009). The best of these reports include third-party validation through an audit.

20.6 World Business Council for Sustainable Development

Just before the 1992 Earth Summit in Rio de Janiero, Brazil, several business leaders sought to give a voice for their sector on sustainability issues in conference discussions. They created a council for business and sustainable development. In 1995, the early council merged with the World Industry Council on the Environment to create the World Business Council for Sustainable Development (WBCSD), with its main office in Geneva, Switzerland, and a second office, open since 2007, in Washington DC, United States. The WBCSD's purposes are to clearly articulate the business case for sustainable development; to encourage members to take a more active leadership role in sustainable development efforts; and to increase outreach to regions where representation is presently weak (World Business Council for Sustainable Development 2009).

WBCSD is an unusual organization in that it was created by business CEOs for business CEOs. It has a membership of more than 200 companies dealing exclusively with business and sustainable development. Members come from at least 35 countries and 20 major industrial sectors, such as cement, electricity utilities, forest products, mining and minerals, transportation and tire manufacturing. WBCSD also has collaborations with 57 national and regional business councils. WBCSD's mission statement is, '…to provide business leadership as a catalyst for change toward sustainable development, and to support the business license to operate, innovate and grow in a world increasingly shaped by sustainable development issues.' WBCSD's policy work is distinguished, as they work to create the right conditions for business to make effective contributions to sustainable human progress and to share leading edge practices among members (World Business Council for Sustainable Development 2009).

So, again, how are we to judge if a business is truly 'walking the talk' or just greenwashing?

After all, to continue as an advanced sustainable society, a mining company will still have to dig holes or extract ore by other means. Questions for mining executives in the near future become does the company mine with the best available technology, restrict its invasive activities when they would irreparably harm an ecosystem, and do they fully reclaim areas to their pre-mined condition, regardless of monetary cost? For bona fide sustainability, there will be full industrial ecology and a human population eventually at, or below, its carrying capacity. We have begun the process of trying to achieve a sustainable paradigm. Having started this journey, it is more likely we will reach that goal sooner rather than later. We have to keep going, lest any catastrophic consequences overwhelm us.

Currently many people, especially in the developing countries see sustainability as a rich peoples' problem. As sustainable practices infuse societies, we expect wider realization that we are all part of the problems and all part of the solutions. Environmental problems created within developed countries and developing countries may be different in scope and scale, nevertheless, we are all on the same planet. Simply put, living unsustainably is everyone's problem. In the United States, one visionary is Van Jones, who works with poorer communities to promote sustainability with an unusual message: Let's create green jobs, rather than wait for green technological solutions.

20.7 The Fourth Quadrant and the Green Collar Economy

Van Jones (2008) has a model of business for sustainability he names 'The Fourth Quadrant.' The model has two axes, a vertical one separating gray problems from green solutions and a horizontal one denoting rich and poor. In the gray half of the model, he has example environmental problems as noticed by wealthy people, such as polar bear extinction and saving whales, whereas poor people are more concerned with local pollution. On the green solutions side, Jones places wealthy solutions such as solar panels and hybrid cars, but poor people cannot afford these solutions, and so their solutions to environmental problems are seen as a need for green jobs and a decrease in environmental injustices. The Fourth Quadrant – where the poor and the green meet – is rarely addressed in environmental messaging.

Jones powerfully calls for development of a Green Collar Economy, creating jobs to attack the environmental problems of the poor. Issues of environmental justice, along with many social and racial issues, can be fixed by making the switch to a Green Collar Economy, Jones argues. Looking into the near-future of the rich-poor separation, Jones (2008) hypothesizes three possible scenarios:

- **Eco-apocalypse**: Keep following business as usual, and this is where we end up. Changing nothing leads to an apocalyptic future with devastating consequences for life, and lifestyles, as we know them.
- **Eco-apartheid**: A minor deviation from the status quo is where privileged people are able to financially overpower their acquisition of limited natural resources,

thus reducing environmental impacts within their own lives, shifting them elsewhere. The poor and underprivileged are left suffering more and more of the consequences directly. Jones notes eco-apartheid is merely a speed bump on the way to eco-apocalypse, since it merely delays a collapse. A green economy has to include the majority of all people, not just a majority of affluent people.
- **Eco-equity**: If we embrace social and ecological interdependency and decide that we need each other and must work together, an eco-equitable future is possible. In this scenario, the wealthy can still afford green technological solutions. The poor become the creators and managers of the new green economy. Green technology requires green technicians.

In a statement that is at once both visionary and practical, Jones (2008) states: 'If you put up solar panels, you're on your way to a professional job with union benefits. This is a "green collar" route out of poverty. When you learn how to double-pane glass to weatherproof a home or install bamboo flooring, you're beginning down a path to a career that is sustainable on both personal and social levels. Jobs like these are the first step on the ladder towards ownership, entrepreneurship and empowerment.' The Green Collar Economy represents a bundle of true win-win solutions.

20.8 Case Study: Walking the Talk in the United States LEED Platinum Certification for the Leopold Center, Baraboo, Wisconsin

In 2007, the Leopold Center was awarded platinum LEED certification (under the second version of LEED) earning 61 points out of 69 available points, possibly making it the greenest building in the United States to date. Planners of the center wanted to use Aldo Leopold's vision of developing positive human relationships with other people and the planet. Through energy efficiency, renewable energy sources and an ongoing commitment to land stewardship, the Leopold Center became the first carbon-neutral building certified by LEED, meaning it accounts for no annual net carbon dioxide emissions. It is also a net zero energy building, meeting all of its energy needs on site. In fact, the building was originally forecast to generate 110% of its own energy needs on an annual basis, although they are still adjusting their expectations as they go. They are still linked into the grid system selling energy into the grid where possible or buying energy from the grid as the need arises! These accomplishments are made despite being in northern Wisconsin's often harsh four-season climate. The center has a roof-mounted photovoltaic solar array on the south side of the main roof. It has geothermal radiant floor heating and cooling thanks to 600 linear feet of 24-inch-diameter cement pipe buried in front of the center. Come winter, heat is supplemented with several fireplaces and stoves using wood harvested from the property. Likewise, much of the wood used in construction of the center came from trees harvested on the property. The wood lot is a legacy, containing many trees planted by Leopold and his family in the days when their shack – famed as the place where *A Sand County Almanac* was written – was the only structure there. The building expresses an aesthetic that is both functional and environmentally sound. Center staff and architects think Aldo Leopold would approve.

Credit: Walking the Talk in the United States: LEED Platinum Certification for the Leopold Center, Baraboo, Wisconsin, United States. Mark Heffron/The Kubala Washatko Architects.

20.9 Case Study: Walking the Talk in Sweden
Corporate Sustainability Reporting by Svenska Cellulosa Aktiebolaget SCA, Sweden

The SCA group shows that business and sustainability can co-exist and benefit each other. For SCA's latest annual and sustainability reports, the forestry and papers products company was awarded an **A+** categorization for adherence to the GRI criteria, including having the reports verified by third-party evaluation. SCA is also a member of the Organisation for Economic Co-operation and Development. SCA stands out as a company with a new vision of corporate social responsibility and goals of ever-more sustainable practices. For instance, they have implemented a new quantifiable carbon reduction target. Fossil fuel emissions are to be reduced by 20% between 2005 and 2020. They have signed the Global Compact (the world's largest voluntary corporate responsibility initiative) and continue to be a major supporter of Forest Stewardship Council certification for forest management. In December 2008, SCA and Statkraft (Europe's largest renewable energy generator) applied for permission to build 455 wind turbines with a total annual capacity to produce 2.4 terawatt hours of electricity. The company is ranked highly by numerous international business watchdogs, such as the 'FTSE4Good' global sustainability index. SCA has been named one of the world's most ethical companies by the Ethisphere Institute, ranked as one of the world's most sustainable companies by the responsible business magazine *Canadian Corporate Knights* using research from social investment firm Innovest, and ranked fifth by the Carbon Disclosure Project's study in the Nordic region.

Credit: Walking the Talk in Sweden: Corporate Sustainability Reporting by Svenska Cellulosa Aktiebolaget SCA, Sweden. Pär Altan, Vice President Media Relations, Corporate Communications.

20.10 Epilogue

Growing green collar jobs in the economy's fourth quadrant, incorporating designs for long-term social and environmental responsibility in industry, and modifying household habits rooted in a family's desire to live their environmental literacy. Any new environmental practice cannot be implemented across society unless communication and education are key factors in reaching out to people. An idea as gigantic, multi-faceted and downright difficult as sustainability will take similarly gigantic, multi-faceted and intricate communication campaigns and educational projects as integral components of the transformation.

Quite possibly the most influential change agent on the planet is the business sector. Undoubtedly, business must be part of any dialogue on sustainable development. Their ability to influence notwithstanding, businesses will not make what consumers will not buy. So, the sovereign consuming public plays the pivotal role in the whole process of a transition to sustainability. Environmental communicators speaking for or about business help consumers and business leaders understand options along the way to a sustainable future. Of types of communications targeted at consumers, advertising is the most pervasive and persuasive form in getting us to buy things we never knew we needed. At the end of the day, though, we decide to buy those products and so hold the decision-making keys to the consumption-driven lifestyle. In Chapter 3, we outlined the contentions of Goleman (2009) that radical transparency about the full life-cycle costs of goods begets high ecological intelligence, assisting us in being wise makers of informed consumer choices. Communicators are the facilitators of this transparency.

The editorial opening the inaugural issue of the *International Journal of Sustainability Communication* (Heinrich et al. 2007) states: 'Alongside sustainability-oriented political regulation and economic self-regulation, communicative, participative and cooperative approaches are seen as essential in harnessing social, economic and ecological complexity, and in facilitating collective transformation processes.' One of the intellectual beauties of humans is the way we keep rethinking issues and keep pressing each other on solving social ills. Whether they carry the modifier 'sustainability' or 'environmental,' communicators will be in the thick of this transformation for a long time.

As we have presented their work here, environmental communicators generate and disseminate planned messages on, about and for improving human–nature interactions. In other words, environmental communicators like to talk big. We hope we have contributed principles and skills to help your big talk produce big impacts.

References and Further Reading

Corporation 20/20 (2008) New principles for corporate design. http://www.corporation2020.org/index.htm. Cited 29 July 2009

Doane D (Fall 2005) The Myth of CSR: The problem with assuming that companies can do well while also doing good is that markets don't really work that way. Stanford Soc Innovation Rev, pp 22–29

References and Further Reading

Edwards AR (2005) The sustainability revolution: Portrait of a paradigm shift. New Society, Gabriola Island, BC

Esty D, Winston A (2006) Green to gold: How smart companies use environmental strategy to innovate, create value, and build competitive advantage. Wiley, New York, NY

Global Reporting Initiative (2009) GRI reports list database 1999–2009. http://www.globalreporting.org/GRIReports/GRIReportsList/. Cited 29 July 2009

Goleman D (2009) Ecological intelligence: How knowing the hidden impacts of what we buy can change everything. Broadway Books, New York, NY

Harwood BB (1997) The healing house: how living in the right house can heal you spiritually, emotionally and physically. Hay House, Carlsbad, CA

Hawken P (1994) The ecology of commerce. Harper Collins, New York, NY

Hawken P, Lovins A, Lovins LH (1999) Natural capitalism: Creating the next industrial revolution. Little Brown & Co, New York, NY

Heinrichs H, Michelsen G, Möller A (2007) Communicating on communication for sustainable development. Int J of Sustain Commun 1:1–2

Jones V (2008) The green collar economy: How one solution can fix our two biggest problems. HarperCollins, New York

Leopold A (1949) A sand county almanac. Oxford University Press, New York, NY

McDonough W (1992) The Hannover principles: Design for sustainability. William McDonough Architects, New York, NY. See also www.mcdonough.com/principles.pdf

McDonough W, Braungart M (2002) Cradle to cradle: Remaking the way we make things. North Point Press, New York, NY

Scheuer CW, Keoleian GA (2002) Evaluation of LEED using life cycle assessment methods. National Institute of Standards and Technology, U.S. Department of Commerce, Gaithersburg, MD

Senge PM (2006) The fifth discipline: The art & practice of the learning organization. Brealey Publishing, London

Senge PM, Smith B, Schley S, Laur J (2008) The necessary revolution: How individuals and organizations are working together to create a sustainable world. Doubleday Publishing, New York, NY

Thomas DF (2004) Toward an understanding of organization place building in communities. Dissertation, Colorado State University, Ft Collins, CO

Thomas DF, Cross JE (2007) Organizations as place builders. J Beh Appl Mgmt 9(1):33–61

Thomas DF, Gaede D, Jurin RR, Connolly LS (2009) Understanding the link between business organizations and construction of community sense of place: the place based network model. Community Dev 39(3):33–45

Townsend AK (2006) Green business: A five-part model for creating an environmentally responsible company. Schiffer

US Green Building Council (2009) LEED version 3. http://www.usgbc.org/LEEDv3. Cited 30 July 2009

Van der Ryn S, Cowan S (2007) Ecological Design, 10 Anvth edn. Island Press, Washinton, DC

World Business Council for Sustainable Development (2009) About the WBCSD. http://www.wbcsd.org. Cited 30 July 2009

Index

A
Abiotic systems, 48
Abstracts, 83, 85
Access to information, 288–289
Accountability, 112, 122
Action skills, 45
Adopting new ideas, 92–94, 100
Adoption theory, 93
Allies, 170
Animals/warm fuzzies, 240–241
Antipathy, 201
Anxiety, 209, 217–221
Apathy, 200–201
Apprehension, 218–219
Argumentation, 41, 57–60
    analogy, 58
    evidence, 57–59
    inference, 55, 57–59
    logic, 58, 59
    metaphor, 58
    presumptions, 57–59
    reasoning, 58–60
    similies, 58
Attitudes, 87, 93–98, 100–102, 106
Audiences, 34, 77, 79–85
    analysis, 79
    externals, 88–89
    focus, 129, 132–133, 136
    influencing emotions, 216–217
    internals, 88–89
Audio clips, 246
Authentic items, 240
Axioms, 3–5

B
Beliefs, 94–102, 106
Belief systems, 95, 96, 102
Body language, 225, 226, 228, 231
Brainstorming, 116
Budgets, 80–83, 85
Building relationships, 167–169
Businesses
    contingent, 293
    contributive, 293
    exploitive, 293
    sustainability, 293–307
    transformational, 293

C
Capacity building, 169–170
Case study
    apples and Alar, 279–280
    bottle bills/beverage container deposits, 71–72
    Center for Environmental and Sustainability Education, 23
    conflict management-Maine, 269–270
    conflict management-Mongolia, 270–271
    converged media, 136–137
    corporate sustainability reporting, 305–306
    cultural context at work, 206
    Egypt's National Strategy, 84
    environmental group formation-Taiwan, 177–178
    environmental radio soap opera-Vietnam, 107–108
    environmental speeches that moved the world, 222–223
    EPA and 9/11, 290–291
    global education project, 123
    an inconvenient truth-Al Gore, 37
    Last Child in the Woods, 61
    Leopold Center-Wisconsin, 304–305
    media relations, 261
    Motorola renew, 158–159

Case study (*cont.*)
  multiple intelligences and learning styles, 190–191
  non-verbals at work, 233–234
  organic food campaign, 290
  personal space artwork, 234
  plant a billion trees, 157–158
  project WILD, 35, 36
  Samsung reclaim, 158–159
  Thirst presentation visuals, 238, 248–251
  Universiti Sains Malaysia (USM), 23
Chalkboards, 241–243, 248
Citizenship, 101–102
Civic agency, 169–170
Coalitions, 169, 170
Communication
  academic, 6, 13
  business, 14, 15
  commercial, 16
  environmental information model, 15–18
  model, 15–19, 111
  perspectives
    constructivist, 30
    critical, 30
    mechanistic, 30
    psychological, 30
    symbolic interaction, 30
    systemic, 30
  risk, 273–280
  science, 4
  technical sector, 16, 17
  theory, 30
Communicator responsibility, 4
Community, 164–166, 170, 177, 178
Community relations, 286
Compassion, 174, 200, 201
Computer images, 246–248
Conflict
  analysis, 263
  anatomy of, 264–266
  benefits of, 268–269
  causes, 264
  communication, 267–268
  escalation, 264, 268
  management, 263–271
  reasons for, 264
  resolution, 263, 264, 266, 267
  unmanaged, 265–266
Consistency
  international, 289
  time, 289
Consumer choice, 106
Consumerism, 90, 105–106
Consumer preference, 105–106

Contextual facts, 55
Controversy, 63
Converged media, 140, 152–157
Convergence
  content, 127–128
  media, 128, 133
Coping styles, 181, 186
Corporate 20/20, 300
Corporate gifts, 289
Corporate social responsibility, myths, 294–295
Corporations, sustainability reporting, 302, 305–306
Cost, 129, 130, 133, 135, 136
Cradle to cradle, 295
Credibility, 273, 276, 278, 279
Crisis communication, 286
Critical thinking, 54–57
Cultural adaptation theories
  assimilation, 196
  conformity, 196
  melting pot, 196
Cultural awareness, 203, 204
Cultural competency, 203, 204
Cultural conflict, 205
Cultural identity, 194, 196
Cultural knowledge, 203
Cultural sensitivity, 203, 205
Cultural suppression, 196–197
Cultural umbrella, 194
Culture
  high context, 197, 206
  low context, 197, 206

D
Data collection
  brainstorming, 116
  case study, 123
  Delphi technique, 117
  expert opinion, 118
  focus groups, 116, 124
  group consensus, 116, 117
  key informants, 115, 118, 124
  nominal group process, 116–117
  professional judgment, 118
  secondary analysis, 117, 124
Data limitations, 278
Decision-making, 123, 166–168, 175
Delphi technique, 117
Dependence, 165–166
Disabilities, 188–189, 202–203
Disputes
  legal remedies, 267

negotiation, 266–267
resolving, 266–267
rule-making, 267
third-parties, 267
Dominant social paradigm, 200

E
Early adopters, 90, 104
Early majority, 90
Earth, 3–7, 9, 10, 13
Eco-apartheid, 303–304
Eco-apocalypse, 303, 304
Eco-equity, 304
Ecological design principles, 297
Education, 165, 170
   adult, 33
   community place, 33
   formal, 31–32
   informal, 32
   non-formal, 32
   sustainable development, 22–24
Electronic media
   facebook, 11
   internet, 41, 42, 48, 50, 53–54
   iPhone, 11
   MySpace, 11
   Wikipedia, 13, 14
Electronics, 240
Emotional intelligence, 173–175
Empathy, 45, 49, 174–175, 200–201, 203–205
Empowerment, 165, 170
Empowerment variables, 101–102
Enneagrams, 185
Entertainment, 127, 130–132
Entry level variables, 101–102
Environmental books, 10
Environmental communication
   definitions
      formal, 14
      informal, 14
   ecological model, 15, 18–19
   goals, 22
   growth, 9 13
   history, 5–9
   human activity, 3
   models, 15–19
      cyclic, 28
Environmental education, 31–35
Environmental interpretation, 31–35
Environmental journals, 6
Environmental responsible behavior, 90, 101–102
Environmental support

basic browns, 91
greenback greens, 91
grousers, 91
sprouts, 91
true-blue greens, 91
Ethics, 4, 23
Ethos, 213
Evaluation
   formative, 81–82, 85
   front end, 81, 85
   influencing factors
      autonomy, 122
      cost, 120
      expertise, 121
      risk of failure, 121
      sampling, 121
      timeliness, 122
      utility, 121
   methods, 112–118
   planning, 122–123
   purposes, 111–112
   summative, 82
   types
      formative, 119
      impact, 120
      outcome, 120
      process, 119
Evaluation plan, 122–123
Expert opinion, 118

F
Facial expressions, 226, 228
Feedforward, 29, 239, 245
Field dependent learners, 186
Field independent learners, 186
Flipcharts, 241–243, 248
Focus groups, 116, 124
Fourth quadrant, 303–304, 306
Fragmentation, 91–92
Framing, 69–71

G
Garden metaphor, 23
Gardner's multiple intelligences, 186–187
Gatekeepers, 115
General public, 88
Gestures, 226, 228
Global reporting initiative (GRI), 302, 305
Goals, 77, 79, 81–84
Government relations, 286
Green at Fifteen, 47, 48
Green collar economy, 303–304

Green to gold, 300–301
Greenwashing, 287–289
Group climate, 167
Group consensus, 116–117
Group existence
 primary, 164
 secondary, 164
 special aspects, 164–166
Groups
 formal, 163
 informal, 163
Groupthink, 168, 169

H
Handling criticism, 289
Handouts, 239, 243–244, 250
Hannover principles, 295–296
Hazard, 273–278
Hegemony, 164
High context, 197, 198, 206
History, 5–9, 19
Human society, 4

I
Implementation, 77, 81, 83, 85
Information
 misconceptions, 50
 misinformation, 50, 55
 providers, 130
 pseudo-science, 50
 reliability, 54
 sharing, 173
 source reliability, 54
Information formats
 brainstorming, 176
 buzz groups, 176
 colloquy, 177
 demonstrations, 175–176
 film, 175–176
 panel discussion, 176–177
 role playing, 176
 speech, 175–176
 symposia, 177
 talk-show format, 177
Innovators, 89, 90, 104
Intelligences
 ecological, 49
 interpersonal, 187
 intrapersonal, 187
 kinesthetic, 187
 linguistic, 187
 mathematical, 187

 musical/rhythmic, 187
 naturalistic, 187
 spatial, 187
Intention to act, 101
Interdependency, 48
Internal communication, 286–287
Interpretive theming, 209–210
Intervention, 165
Interviews, 112, 114–116, 124
Inverted pyramid format, 257–258, 260
Issues, 63–72
 analysis, 63–66, 68, 71–72
 creation of, 66
 dissection
  bias, 67
  glamorization, 67
  personalization, 67
  sensationalization, 67
  simplification, 67
 global, 68–69
 local, 68–69
 regional, 68–69

K
Key informants, 115, 118, 124
Kinesics
 body language, 226
 facial expressions, 226
 gestures, 226
 physical movement, 226
Knowledge, filter, 41, 43–47, 49–55

L
Laggards, 90, 104
Land ethic, 295
Language problems, 42, 46
Late majority, 90, 100
Leadership, 165–166, 168, 170, 171, 173
Leadership in energy and environmental
  design (LEED), 298, 304–305
Learners
 characteristics, 184
 field dependent, 184
 field independent, 184
Learning styles
 aural, 188
 haptic, 188
 kinesthetic, 188
 olfactory, 188
 print, 188
 visual, 188
Legislation, 50, 61

Literacy, 41–61
  ecological, 48–50
  environmental, 41–61
    functional, 45
    measuring, 46–48
    nominal, 45
    operational, 46, 60
  primary, 52
  science, 41, 43–45
  secondary, 52–53, 61
Litigation, 49, 267
Locus of control (LOC), 99–102
Logical idealism, 200
Low context, 197, 198, 206
Loyalty, 91–92

M
Macroculture, 194–197, 199
Management
  chaos theory, 171
  contingency theory, 171
  systems theory, 172–173
  theory, 171–173
Marketing, 283–291
Mark, T., 29
Maslow's hierarchy, 104
Mass media
  characteristics, 129–135
  purpose, 129–130
  risk, 277
Measurements, 50, 56–57
Media, 80–83, 85
  abstracts and executive summaries, 143
  blogs (web logs), 151
  channel, 129, 133
  direct mail, 146–147
  email, 150–151
  films, 148
  information sheets, 145–146
  interpretive talks/presentations, 147–148
  letter writing, 141–142
  mobile device messaging, 151
  modern media, 149–152
  newsletters, 147
  news releases
    mechanisms, 140–141
    problems, 140–141
  options/choices, 80, 81
  podcast tour guides, 154
  public service announcements (PSAs), 143–145
  purpose, 129–130
  reader responses, 153–154
  relations, 285–286
  role, 254
  science writing, 146
  social networking, 152
  streaming events, 156
  traditional media, 140–148
  unusual media, 157
  viral marketing, 154–155
  webinars, 155–156
  World Wide Web (WWW) sites, 149–150
Mediation, 267
Mediator, 32
Membership trail, 288
Memes, 98–99
Mental models, 173
Message development, 80
Message elaboration, 30–31
Metaphors, 231
Microculture, 194–197, 199, 202
Models, 240
Money trail, 288
Motivation, 101–105, 163, 164, 167, 173–175
  models, 103–104
  needs, 100, 103–104
Multiple intelligences, 186–191
Myers-Briggs
  extrovert-introvert, 183
  judger-perceiver, 183
  sensor-intuitive, 183
  thinker-feeler, 183, 184

N
Nature deficit disorder, 49, 61
New environmental paradigm, 200
New media, 8, 128, 129, 149–152, 156, 157
News accuracy
  distortion, 254
  engineers, 259–260
  environment, 256–257
  inaccuracy, 256
  oversimplification, 254
  science, 255–256
  scientists, 259–260
  sensationalism, 254
News media, options, 258–259
News process, 253–254
News questions, who, what, where, why, when, and how?, 257
News releases, 257–258
News reporting format, 257–258
No Child Left Inside, 49–50, 61
Nominal group process, 116–117

Non-verbals
   consistency, 232–233
   cultural implications, 231–232
Numeracy, 41–43, 45

O

Objectives, 77, 79–85
Observation, 114–115, 124
Openness, 166
Opinion, 87, 90, 91, 93–98, 100, 102, 106
Opinion leaders, 115, 124
Organizations
   environmental agencies, 17
   governmental, 17
   non-governmental, 17–18
Outrage
   decrease, 276
   increase, 276
Overhead projection, transparencies, 244–245
Ownership variables, 101–102

P

Paralanguage, 229
Personality types, enneagrams, 185
Personal mastery, 173
Personal space
   intimate, 227
   public, 227
   social, 227
Persuasion, 8–9, 14, 22, 130, 131
Philanthropy, 289
Photographic slides, 245
Physical movement, 226
Placement, 284
Planning, 77–85
Planning process, 77–83
Platinum rule, 205
Players, 63, 65
Pluralism, 197
Political function, 112, 121
Positions, 63–68, 70, 71
Powerpoint, do's and don'ts, 249–250
Presentations
   conclusion, 212
   delivery, 212–217
   introduction, 211
   main body, 211
   mannerisms, 215–216
   posture, 215–216
   structure, 209
Presentation structure, 209
Presuppositions, examples, 230–231

Price, 284
Problems, 63–67, 69, 70
Problem statements, 78–79, 81–83
Product, 284
Professional judgment, 118
Promotion, 284–285
Propaganda, 67, 70, 287
Proxemics, personal space, 226–227
Public relations, 285–287
   risk, 277, 280
Public speaking
   fears
      conspicuousness, 220
      dissimilarity, 220
      history, 220
      novelty, 220
      subordinate status, 220

Q

Quad stack, 20–22, 293, 295, 298, 302
Qualitative techniques, 118–119
Quality of life, 20–23
Quantitative techniques, 118–119

R

Reconstruction, 70
Regulatory paper trail, 288
Relationship management, 175
Relaxation techniques, 220–221
Reporting
   limitations, 256–257
   public affairs, 8
Reporting constraints
   advertiser pressure, 255
   crisis orientation, 255
   editorial pressure, 255
   lack of knowledge, 255
   lack of sources, 255
   the scoop, 255
   short deadlines, 255
Research
   correlational, 56
   experimental, 56
Resolving disputes, 266–267
Responsible behavior, 90, 101–102, 106
Responsible decision-making, 175
Revelation, 165, 168
Risk
   assessment, 274, 277
   consequences, 274
   evaluation, 274
   management, 274, 275
   outrage, 274–275

Index 309

Risk acceptance
    acts of God, 276
    choice, 275
    environmental justice, 276
    financial burden, 275–276
    news reports, 276
    trust, 276

S
Sanborn principles, 296–297
Satir modes
    blamer, 182
    computer, 182
    distracter, 182
    leveler, 182
    placater, 182
Schoenfeld, C.A., 5–7, 12, 14, 19, 27
Science
    frontier, 52
    method, 43, 51
    textbook, 53
    vocabulary, 43
Secondary analysis, 117, 124
Segmentation, 89–92, 104
Selectivity, 91–92
Self-awareness, 173, 175
Self management, 175
Self-regulation, 173
Semiotics, symbolism, 227–229
Sense of place
    businesses, 298–299
    community place, 20
    cultural place, 20
    economic place, 20
    place attachment, 19
    political place, 20
    social place, 20
Shared vision, 170, 173
Situational factors, 97–98
Skeletons in closet, 288
Social awareness, 175
Social intelligence, 173, 174
Social marketing, 283–285, 289
Societal discourse
    adoption, 92–94
    awareness, 93
    evaluation, 93–94
    interest, 93
    trial behavior, 94
Society, 4, 6, 8, 14, 15, 20–22
Sociotyping, 201, 202
Solutions, 63–65, 67, 69
Sourcewatch, 288
Speaker apprehension test, 218–219

Speaker types, 216
Spiral of unmanaged conflict, 265–266
Stakeholders, 65, 68, 69
Stances, 63, 65–67
Standard of living, 21
Statistics, 41, 43, 56–57
Stereotyping, 201, 202
Strategies, 65, 66
Subtyping, 201, 202
Surveys, 112–114, 124
Sustainability, 19–24
    businesses, 293–307
Sustainable development, 293, 296, 302–303
Symbolism, haptics, 228
Sympathy, 201, 203, 205
Systemic thinking, 301
Systems theory
    mental models, 173
    personal mastery, 173
    shared vision, 173
    team learning, 173
Systems thinking, 172, 173

T
Target audiences, 80, 83–85, 87, 88, 93, 95,
        100, 101, 103, 108
Team building
    negative, 167
    positive, 166
Team learning, 173
Teams, self-directed, 173
Thematic statements, 210
Themes, 210
Theming, green, 287
Third party technique, 288
Time constraints, 10
Timelines, 80–83
Timeliness
    frequency, 134
    longevity, 134, 135
Triple bottom line, 20–22, 293, 295, 301, 302
Trust, 174, 273, 276, 278

U
Uncertainty, 277–278
Understanding, 3–24

V
Value descriptors
    aesthetic, 68
    cultural, 68
    ecological, 68

Value descriptors (*cont.*)
  economic, 68
  educational, 68
  egocentric, 68
  legal, 68
  recreational, 68
  religious, 68
  social, 68
  spiritual, 68
Values, 64, 65, 67, 68, 70, 71, 94–98, 100, 102, 105, 106
  aesthetic, 263
  ecological, 263
  economic, 263
  educational, 263
  environmental, 263
  ethical, 263
  spiritual, 263
Verbal delivery, 212–217
Verbals
  inflection, 214
  pitch, 213–214
  projection, 214–215

Video clips, 246, 250
Visual aids, basics, 237–239
Visual usage techniques, 250–251
Vocal qualities, 213–215

W

Walking the Talk, 293–307
Whiteboards, 241–244, 248
Win-win, 264, 266–269
World Business Council, 302–303
Worldviews, 94–98, 100, 101
  African, 199
  Asian, 199
  Euro-American, 199
Writing
  nature, 7
  outdoors, 7
  pioneers, 7
  recreation, 7
  science, 7–8
  travel, 7

Breinigsville, PA USA
31 August 2010
244580BV00012B/30/P